LIFT,
THRUST & DRAG
—a primer of modern flying

Other TAB books by the author:

LIFT, THRUST & DRAG
—a primer of modern flying

by Paul Garrison

MODERN AVIATION SERIES

TAB BOOKS Inc.

BLUE RIDGE SUMMIT, PA. 17214

FIRST EDITION

FIRST PRINTING

Copyright © 1981 by TAB BOOKS Inc.

Printed in the United States of America

Reproduction or publication of the content in any manner, without express permission of the publisher, is prohibited. No liability is assumed with respect to the use of the information herein.

Library of Congress Cataloging in Publication Data

Garrison, Paul.
    Lift, thrust & drag.

    Bibliography: p.
    Includes index.
    "TAB book #2309."
    1. Aerodynamics.   2. Aeronautical instruments.
3. Electronics in aeronautics. 4. Aids to air navigation. I.
Title. II. Title: Lift, thrust, and drag.
TL570.G26   629.13   81-9255
ISBN 0-8306-2309-4 (pbk.)   AACR2

Contents

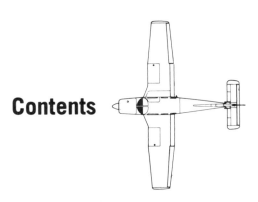

Introduction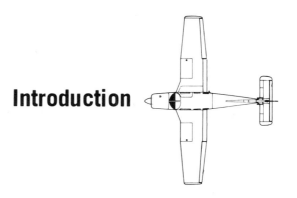

It is my intention to provide with this book a primer to modern flight, to explain in some detail what flying is all about in today's aircraft and airspace. Other books have, of course, been written on this subject but few, if any, seem to have taken into account the enormous changes which have been taking place with reference to aircraft sophistication and airspace regulations, not to mention instrumentation. The fact is that a Rip van Winkle-type pilot, having learned to fly twenty or thirty years ago, then having slept his long sleep to be awakened today, would be faced with nearly insurmountable difficulties in picking up where he left off.

And no wonder.

In the less than 80 years since that day when Orville and Wilbur succeeded in proving the feasibility of powered flight by staying airborne for 57 seconds at an airspeed of 31 miles per hour, powered by an engine which developed all of 12 horsepower, we have built piston engines with horsepower ratings in the thousands, have advanced to turbine power which enabled us to break the sound barrier, and eventually to rockets which took us all the way to the moon. No other technology in history has advanced so far in such a short span of time as has flight.

But still today the design of all aircraft (not including space vehicles) is based on the simple concepts of *lift*, *thrust* and *drag*, concepts which the Wright Brothers were the first to put into practical perspective, combining them with what little they guessed rather than clearly understood of that important ingredient which we refer to today as *angle of attack*.

Flight is possible only when thrust exceeds drag and when lift is provided by the fact that the air, the so-called relative wind, meets the lifting surfaces, the airfoils, at the correct angle, the *angle of attack*. Strangely enough, a large proportion of today's pilot population has no clearer idea of what that angle of attack is all about than did those two courageous experimenters on Kill Devil Hill. Most, in fact, know *less*, not having had to spend years in trial and error to develop a machine which was capable of flight.

The fact is that a huge percentage of the light-aircraft accidents today can be traced directly to this lack of understanding and appreciation of what makes an airplane behave the way it does. Any pilot who fully understands what the angle of attack is all about could not possibly get himself into a situation in which he inadvertantly stalls his airplane, ending up in a low-altitude spin—and dead.

For this reason, the first part of this book will be devoted primarily to what makes an airplane fly and why, under certain circumstances, it will suddenly stop flying. Only a complete understanding of the frequently illogical-seeming behavior of the airplane and its controls can produce a safe and therefore eventually old pilot.

Another part of the book will concern itself with what is commonly referred to as air-data instrumentation. Air-data instruments are the basic instruments commonly found in all aircraft: The airspeed indicator, the altimeter, the compass, the vertical-speed indicator, needle and ball, and so on. Once flight has become more or less instinctive, an understanding of the peculiarities of these instruments and the ability to read them correctly at a glance is the next most important step.

A third part then covers navigation and communication and the related instruments, known in aviation as *avionics*. The type of avionics found in virtually any aircraft today was virtually unknown thirty years ago. Then, flying by dead reckoning, pilotage, and the highly overrated "seat-of-the-pants" was the rule rather than the exception. Today it would be difficult, indeed, to find a pilot embarking on a flight who will not use his avionics equipment for navigation and, when called upon, for communication. We live in an electronic age, and without a complete understanding of how best to use the electronics in our cockpit, much of the airspace in the United States would be off limits.

And then I intend to touch on some of the more sophisticated systems, those convenience items which are available at a price and which, while not necessities, make flying easier and more enjoyable while at the same time improving the safety and efficiency of both airplane and pilot.

This book is *not* intended to take the place of a flight instructor. No written word can. What it *is* intended to do is to augment what is taught at ground school and in the cockpits of training aircraft. Flight instructors have historically varied in quality from quite bad to excellent. The excellent ones are the true teachers among them, those who love to teach and who look at their occupation as their life's work despite the generally low level of income that can be derived from teaching. The other ones are usually those who consider instructing a means of accumulating the time necessary in order to eventually qualify for a job as pilot with a corporation or airline. They, too, are usually perfectly good pilots, but more often than not their ability to teach others may be quite limited.

Without wishing to generalize, my own experience with flight instructors has been mostly with the second variety, and looking back on close to three decades and nearly 5,000 hours of flying, I realize time and again that whatever it is that I know about the subject, I somehow managed to teach myself. And most veteran pilots would probably agree that practical experience is by far the best teacher. Theory in the classroom or the words in articles or a book are all very nice, but they cannot make a pilot. Dual instruction is important to a point, but it would literally take hundreds upon hundreds of hours of such instruction to prepare the student for all possible eventualities. The fact is that no two flights are ever exactly alike, and only a combination of theory (leading to understanding) and practice (leading to experience) can ever hope to produce the kind of pilot who is prepared to face whatever he may encounter.

Chapter 1
What Makes
an Airplane Fly?

You ask this question of the average non-pilot and he'll probably say, "The wings, naturally." And to a degree he is right. The wings do have something to do with permitting the airplane to fly, but left to their own devices, they'd simply plummet to earth. Another, asked the same question, may say, "The engine." But if an engine makes the airplane fly, how come gliders and sailplanes stay up there for hours on end with no problem at all? But if it's neither the wings nor the engine, what is it?

The fact is that the ability of an airplane, powered or not, to stay aloft is the result of the interaction of a natural phenomenon, namely the air, with a man-made contraption, namely the airplane (Fig. 1-1). It takes both, the air and the machine, to act upon one another in accordance with certain principles, the principles of aerodynamics, in order to permit flight to take place. The horizontal stabilizer (or horizontal tail surface) must be kept in movement through the air at a given minimum speed and at an angle to the airflow which is neither too shallow nor too steep, in order for flight to result.

By now we have arrived at the belief, though not necessarily the understanding, that a combination of a natural substance—air—and a man-made substance—the horizontal surfaces of an aircraft—and a dynamic force—movement—is capable of producing flight. But how, exactly, does all this work?

Tens of thousands of words have been written about the fact that because of the curvature of an airfoil (read wing), the air above

Fig. 1-1. What is it that makes an airplane fly?

that airfoil (wing) is less dense than the air below it, thus causing a suction effect which keeps the airfoil (wing) from dropping. While there is *some* scientific substance to this argument, it is really a bunch of nonsense as far as the pilot is concerned. To the pilot, it should make *no difference* whether the wing is pushed up by compression of air below the wing or by suction produced by a lack of air above the wing. If *this* is what he is thinking about when flying, he is bound to get himself into trouble.

The real secret of flying consists of just three words: *Angle of attack*. It makes no difference what kind of airplane is involved or what shape the wings are, it'll only fly as long as an acceptable angle of attack is maintained. Screw up the angle of attack and it may turn out to be the last thing you ever did.

Angle of Attack

All right, already. So what's this famous angle of attack all about?

The angle of attack is the angle at which the air strikes any airfoil. Now what the heck does that mean? In the belief that a picture is worth a thousand words, let's look at Figs. 1-2 through 1-4 before reading on. Since the actual shape of the airfoil is really of no consequence when we're trying to understand the effect of the angle of attack, I have used a simple barndoor-shaped slab in these illustrations. In each instance the solid black "airfoil" is thought to be moving from right to left, meaning that the movement of the air relative to the airfoil is left to right (⟶). In Fig. 1-2, the air (or, to be technical, the *relative wind*) strikes the airfoil at an angle which is sufficiently shallow to permit the air to continue to flow

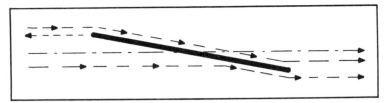

Fig. 1-2. At a small angle of attack the air flows smoothly along the surfaces of the moving object.

smoothly. This smooth airflow results in lift. In Fig. 1-3, the angle of attack is steepened to a point which causes the airflow to break up. No lift is created. In Fig. 1-4 we show what would happen if the airfoil strikes the relative wind at a 90-degree angle (not a very likely occurrance). All of these illustrations are drastically simplified, but they do tend to illustrate what the concept of angle of attack is all about.

What all this really means is that as long as the airfoil strikes the relative wind at an acceptable angle of attack (and with sufficient speed, of course), it cannot possibly do anything other than produce lift. But as soon as this angle of attack is increased beyond a certain acceptable limit, lift is lost and the airfoil (and the airplane which is attached to it) will start to drop toward the ground. In other words, flying is really quite simple. All we have to do is to retain an acceptable angle of attack. Big deal! I can only say that I wish that it would be as simple as it sounds.

At this point you might say that if the angle of attack is that important, logically there should be an instrument in the cockpit which tells the pilot what his angle of attack is at any given

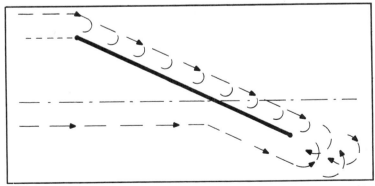

Fig. 1-3. When the angle of attack is increased beyond a certain limit, the air becomes disturbed and breaks away from the surfaces of the moving object.

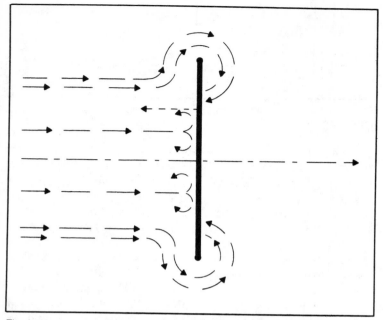

Fig. 1-4. At this admittedly extreme angle there is only drag and no lift.

moment. And you'd be right in thinking that this should be so, except it *isn't*. Such instruments do exist and they are not even particularly expensive. But in practice, light airplanes equipped with angle-of-attack indicators are few and far between. Thus, with no such instrument to guide us, we have to develop a sort of sixth sense that tells us when we're about to exceed the maximum permissible angle of attack. But more about that later.

Airspeed

First, let's look at something else, namely *speed*. In order for the air, the relative wind, to strike the airfoil at all, there must be movement. If there is no relative wind there is no angle of attack and, in turn, no lift. Now, in order to produce speed we need an engine. Right? *Wrong*. Sailplanes fly perfectly well and they have no engine. What this tells us is that speed can be achieved through either of two means. If we have an engine, fine; it will provide the thrust which results in speed. If we don't have an engine, the gravity pull of the earth can take its place (Figs. 1-5, 1-6). All we have to do is to put the nose of the airplane down relative to the air surrounding it; speed (and, in turn, lift) will result.

But what kind of speed are we talking about? The speed which is a necessary prerequisite for lift is *airspeed*; in other words, the speed with which the airplane is moving *through the air*. It makes absolutely *no* difference to the airplane's ability to fly at what speed, if any, it is moving across the ground. If the minimum flying speed (known as *stall speed*) of an airplane is, say, 50 miles per hour and if the wind is blowing at 60 miles per hour, the airplane may actually be standing still or even moving *backwards* with reference to the ground, and it will still be flying. Ground speed is important if we want to know how long it is going to take us to get from here to there. But as far as actual flight is concerned, it is meaningless. See Figs. 1-7 through 1-9.

Drag

In addition to the angle of attack and the means of attaining a certain amount of airspeed, there is a third ingredient which influences the resulting lift and the amount of thrust needed in order to produce adequate airspeed. This third ingredient is the negative effect of *drag*. Air is not void space. It has quite considerable weight and mass, and any object being moved through the air will experience a degree of resistance. In aviation, this resistance is described as *drag*. Any airplane, no matter how clean, produces two types of drag, known as *parasite* drag and *induced* drag.

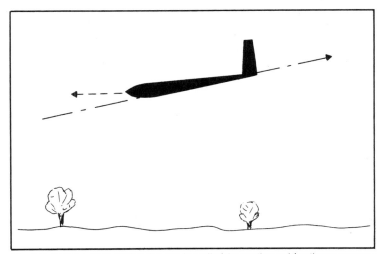

Fig. 1-5. If we have no engine the gravity pull of the earth provides the necessary thrust.

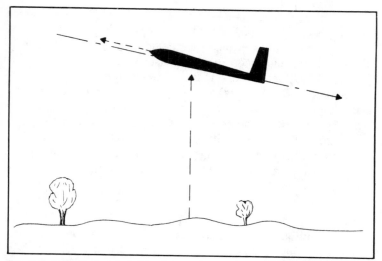

Fig. 1-6. If the air is rising, the glider can rise with it while still maintaining a sufficient downward movement within the air to result in thrust.

Parasite drag is the resistance produced by any surface or portion of an airplane which does not contribute to lift. Its effect increases with increased airspeed. *Induced* drag is that part of the resistance which results when lift is produced. Remembering the difference between the two is of little practical value unless you plan to design and build your own airplane. As a matter of fact, an awareness of the existence of drag is of no importance to us as pilots, because any airplane we fly will have been designed to provide adequate thrust to overcome the drag. An understanding of its effect does become important in relation to certain of the control

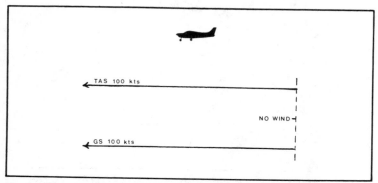

Fig. 1-7. With no wind, true airspeed and ground speed are the same.

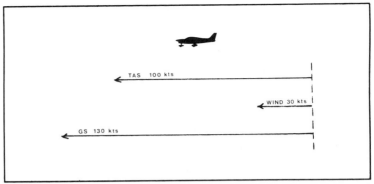

Fig. 1-8. With a 30-knot tailwind, a 100-knot true airspeed will result in a ground speed of 130 knots.

surfaces, but this will be discussed in more detail in the appropriate chapters.

Airfoils

We've been using the term *airfoil*. There are hundreds, possibly thousands of different shapes of airfoil and the science of producing just the right airfoil for varying aircraft and speed conditions is complicated indeed. In fact, the Wright Brothers experimented with over 200 different shapes before deciding on the one they used—which was actually not very good, but it did do the trick. From the point of view of actually flying an airplane, an understanding of the mysteries of airfoil design is of no practical value. We simply have to assume that the manufacturer knew what

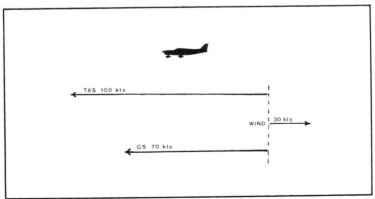

Fig. 1-9. With a 30-knot headwind, a true airspeed of 100 knots will result in a ground speed of only 70 knots.

he was doing, because even if he didn't, there is no way for us to rectify whatever minor errors in his calculations may have found their way into the airfoils on the airplane we are flying.

And there are a lot of places where such minor errors can actually occur. Whenever we talk about airfoils, we automatically think of the wings, because they are the largest and most obvious. But there are a lot of other airfoils on an airplane. The horizontal and vertical stabilizers are airfoils, and so are the ailerons, certain types of flaps, the rudder and elevator surfaces, and even the blades of the propeller (Fig. 1-10). Poor design of any one of these can result in a slight degradation of the flying qualities of an airplane.

Let's just quickly look at the fixed airfoils other than the wings and examine their part in permitting an airplane to fly. First, the horizontal stabilizer, those small wings attached to either side of the tail. If the wings provide the necessary lift to keep the main portion of the fuselage aloft, then it would seem logical that the small wings at the rear are for the purpose of keeping the tail from dropping. Except this is not the case. On most airplanes the engine, being in front and ahead of the center of gravity, represents so much weight that it might actually tend to force the nose of the airplane down, causing the tail to rise. Thus the horizontal stabilizer on most powered aircraft is actually a *negative* airfoil, meaning that it produces a small amount of negative lift, thus preventing the tail from rising above the level of the rest of the aircraft. (This is not necessarily true of *all* aircraft, but it is true of most of the ones we're likely to be flying.) In technical terms, whatever the actual shape of the airfoil of the horizontal stabilizer, its purpose is to provide the aircraft with *pitch stability* (pitch being the movement of an aircraft with reference to its lateral axis, an imaginary line going at right angles through the center of gravity of the aircraft).

Torque Effect

The vertical stabilizer serves a similar purpose with reference to the directional stability of the airplane, but in most powered aircraft (except twins equipped with counter-rotating propellers) it must serve the additional function of overcoming or, at least, minimizing the so-called *torque effect*. This is kind of complicated to explain, and I'm not at all certain that a clear understanding of this torque effect is of particular importance, but since a thorough understanding of what makes an airplane work

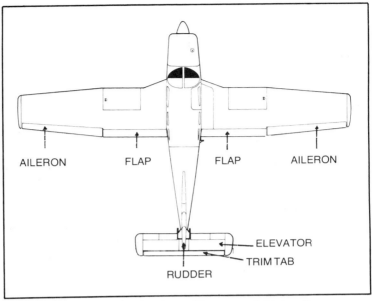

Fig. 1-10. The movable control surfaces of the average light aircraft.

makes a pilot a better pilot, let's look at it for a few minutes in detail.

The most popular explanation of torque is that because the engine turns the propeller one way it will, according to the law of physics that any action produces a comparable reaction, try to roll the airplane to which it is firmly attached in the opposite direction. While this *is* true in theory, it is not actually noticeable in practice. The wings of the airplane are so large in comparison to the size of the propeller blades that they provide the aircraft with sufficient inherent stability to overcome this rolling tendency. But anyone who has ever flown an airplane knows that there *is* a torque effect which causes the airplane to want to turn to the left during takeoff and full-power low-speed climb, while trying to turn it to the right during power-off descents. But this tendency is not the result of what has been described above, but rather of the way in which the airstream produced by the rotating propeller hits (or does *not* hit) the vertical stabilizer.

As a matter of fact, the term "torque" is a misnomer in this instance. What actually happens is that the air being forced backwards by the propeller moves across the aircraft in a spiraling motion. The effect is slight and could probably not be seen if the air were somehow made visible, but it is sufficient to strike the left side of the vertical stabilizer with greater force than it does the right side, thus trying to push the tail to the right and, in turn, producing a tendency for the nose to want to turn to the left. In order to counteract this effect, the vertical stabilizer is installed at a minute angle with the front end being slightly left of center and the rear end being an equally unnoticeable amount to the right of center. This helps to minimize the force with which the airflow hits the sides of the vertical stabilizer while at the same time acting like a bit of right rudder.

As I said before, this canting of the vertical stabilizer is so slight as to be undetectable with the naked eye, the amount being calculated to keep the airplane from yawing when in straight and level cruise at normal cruise power. The tendency of the airplane to yaw to the left during takeoff* and to the right during a power-off approach can now easily be explained. During takeoff and full-power climb we use more than the normal power, thus produce a greater-than-normal spiraling motion of the air across the airplane. This is further exacerbated by the fact that we are operating at very low airspeeds. The only way to keep the airplane moving along a straight line is to use right rudder. During the power-off approach the opposite occurs. There is no spiraling slipstream to act upon the vertical stabilizer and it, being offset to allow for that spiraling airstream, now acts as right rudder, trying to turn the nose of the airplane to the right. Here, then, we have to apply a degree of left rudder in order to counteract that tendency.

The beginning flight student might wonder how he will ever be able to remember to do the right thing at the right time. He needn't worry. It very soon becomes a completely automatic reaction, one which requires no conscious thought on the part of the pilot.

So much, then, for what makes an airplane fly. *Sufficient thrust to overcome drag coupled with the right angle of attack produces lift while the secondary fixed airfoils help to create an acceptable degree of directional and longitudinal stability.* But the angle of attack is too important to take a chance on having it only partially understood. So let's take one closer look at it.

---

*Another phenomenon known as *gyroscopic precession* may also contribute to left yaw during takeoff and certain other flight regimes.

Chapter 2
Angle of Attack—The Heart of the Matter

To repeat, the angle of attack is *the angle at which an airfoil strikes the relative wind.* Okay, but what does that really mean? For some reason, it is relatively easy to visualize when we think in terms of level flight. But in level flight, at cruise, the angle of attack is something we don't have to worry about. The time when it becomes important is during some of the more marginal flight profiles, such as slow flight, climb and descent. And it is during these times when trying to visualize the actual angle of attack becomes complicated.

Relative Wind

Instinct wants to tell us that when the nose of the airplane is pointed in a given direction, then that is the direction in which we are flying, ergo, the relative wind comes from the direction toward which the nose is pointing. *Not so.* Whenever we are operating at lower than normal airspeeds, the nose of the airplane will point several degrees *above* the direction in which we are actually moving through the air. Thus the angle of attack will be several degrees greater than it is during level cruise (Fig. 2-1).

Another mistaken idea often found bouncing around in the heads of student pilots is that there is some relation between the angle of attack and the angle of the nose of the airplane relative to the horizon. *No such relation exists.* During a zoom or while climbing a really powerful airplane, the nose may be pointing high

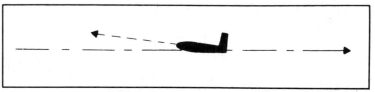

Fig. 2-1. When we are operating at lower than normal airspeeds, the nose of the airplane will point up and the angle of attack is greater than in cruise.

above the horizon and still the angle of attack will be quite small (Fig. 2-2). By contrast, the nose may be below the horizon and the angle of attack be quite extreme, the worst example of this being the condition of a spin.

It can't be stated often enough: The angle of attack is the angle at which the air strikes the wing. When this air, the relative wind, and the wing are perfectly aligned then there is *no* angle of attack and the wing cannot develop any lift. (Never mind all that theory about the air moving faster across the upper curved surface of the wing than it does across the lower surface, thus being thinner on top, resulting in suction which holds the wing up. True, there is some substance to this, but it only tends to confuse the issue, and really is of no practical use to the pilot.) The relative wind always strikes the wing at an angle from below, no matter how slight.

The Stall

At high speed—that is, high *airspeed*, not ground speed—the force of the relative wind is considerable, requiring only a very small angle of attack in order to produce the necessary lift (Fig. 2-3). The lower the airspeed, the less the force of the relative wind, requiring ever-greater angles of attack in order to keep the airplane aloft. Up to a point! An increase in the angle of attack results in an increase in lift *until*, and this is most important, until the angle becomes so great that the airstream flowing across the wing becomes disturbed. This severe disturbance of the airflow causes a relatively sudden and complete loss of lift resulting in what is known as a *stall*. Stall means that the airplane is moving through the air at a speed which would require an angle of attack so great that it exceeds its capability of producing lift. (Stalls are usually thought of as being the result of excessively slow flight. This is not always the case. There are also high-speed stalls.)

The point of all this beating the drums for the angle of attack is that as long as an acceptable angle of attack is maintained, it is physically impossible for the pilot to get into trouble, excepting

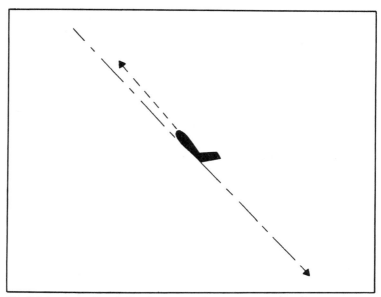

Fig. 2-2. In a zoom the airplane's nose may be quite high, but as long as speed is maintained, the angle of attack is no greater than in cruise.

such things as mid-air collisions or flying into the side of a mountain. But how are we to know when the acceptable angle of attack is about to change to an unacceptable angle?

In airplanes equipped with an angle-of-attack indicator, this is simplicity incarnate. An angle-of-attack indicator, as far as the pilot is concerned, is simply a needle which moves along a scale, marked "slow" at one end and "fast" at the other (Fig. 2-4). When the needle is at the fast end, the angle of attack is small, such as is normal during cruise. When it is halfway between the two, it indicates an angle of attack which is just about right for best rate of climb or a normal approach. When it's on the slow end, the angle of

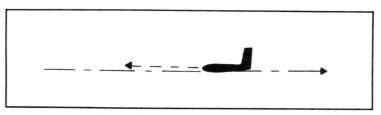

Fig. 2-3. At high airspeed, the direction of flight of the aircraft and the direction of the relative wind are virtually parallel (though in opposite directions) resulting in a very small angle of attack.

attack is so great that any further increase would result in a stall. Thus, when the pilot sees the needles creeping toward the slow side of the scale, his automatic reaction should be to lower the nose. This stops the needle movement and prevents the angle of attack from exceeding safe limits.

There are many times in the life of a pilot, novice and veteran alike, when this kind of easy-to-interpret instrumentation is an extremely helpful adjunct, and it has always struck me as strange, if not actually criminal, that an angle-of-attack indicator is not provided as standard (or better still, *mandatory*) equipment in light aircraft. Many a stall-spin fatality might have been avoided if the pilot had been able to judge his angle of attack with certainty the moment *before* he got himself into trouble.

Aircraft Feel

Without such an instrument, all we have to rely on is the so-called "feel" of the airplane. As long as the angle of attack is kept within reason, the airplane will behave more or less normally, responding to the controls the way it should. As the angle of attack increases, this ability to respond to the controls diminishes, and we think of the airplane as "mushing" through the air. At the point of a stall, the point when the angle of attack has become too great to continue to produce lift, control response is at a minimum, and the only way to regain it is to put the nose of the airplane down, getting the angle of attack back to an acceptable level. It is the one time during flight when we *must* depend on the so-called seat-of-the-pants, despite that fact that at many other times this same seat-of-the-pants can deceive us and get us into serious trouble.

The one saving grace about airplanes is that they *want* to fly, and to maintain a lift-producing angle of attack. At any time at which the pilot takes his hands off the controls, the airplane will seek out and eventually regain the speed (and in turn, the angle of attack) for which it is trimmed. No airplane will ever get into a stall (or a subsequent spin) without some type of drastic action by the pilot. It simply cannot happen. The trouble is that the tendency to exceed the acceptable limits of the angle of attack occurs usually close to the ground, putting the pilot into a quandry. He knows (or should know) that the only way to keep the angle of attack within limits is to put the nose down. But *down there* is the ground and every bone in his body cries out to have him pull up rather than push down, thus making an eventual stall inevitable. And then, when the stall does come, the airplane, in a willful attempt to regain flying speed,

ONE EIGHTY-THIRD NCO CLUB
Capital Airport   Springfield, IL

SGT Richard H. Shanner

Is a member in good standing

Dues paid to 31 Dec 83

Card Number   00489

Signature   _Robert M. Stewart_

Secretary-Treas.

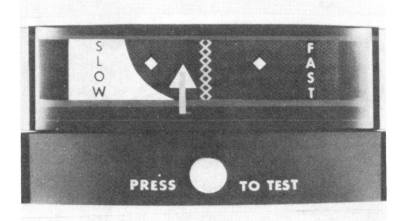

Fig. 2-4. The cockpit display is an angle-of-attack indicator (courtesy of Safe Flight Instrument Corp).

drops its nose because it has no way of knowing that there is solid ground only a few feet away. The result, at best, is a bent propeller and busted nosewheel. At worst . . . well, never mind.

In some of the chapters on flying technique we'll go into this subject in somewhat greater detail. Suffice it for now that understanding and respecting the angle of attack is the basis for _____ in fact, all there is to the theory of flight in a

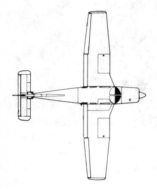

Chapter 3
The
Components
of the Airplane

All airplanes consist of two categories of components: *fixed* and *movable*. The fixed components are the *fuselage*, the *wings* and the *horizontal* and *vertical stabilizers*. They are the basic structure which permits the airplane to fly. They carry the passengers or freight and contain the engine and fuel. The movable components are the *ailerons, rudder, elevator, flaps, trim tabs,* and in some cases *spoilers*. Their purpose is to make the airplane controllable, to permit it to climb or descend, fly fast or slow, and to make turns. The propeller, too, might be thought of as one of the movable components, but since gliders, too, are airplanes, a discussion of the propeller might be more appropriate in conjunction with a discussion of the engine.

The Fuselage

The *fuselage* is the main body of the airplane consisting of the *engine compartment*, the *cabin*, and the *tail cone*. While, in terms of volume, it is by far the largest portion of the whole airplane, it is the only one which does not measurably contribute to its flying capability. (I say measurably, because, depending on its shape, it may actually produce a minute amount of lift.) On the contrary, it is the primary generator of parasite drag which is the reason why the cabins of most light aircraft are small and somewhat cramped (in comparison to, say, a Cadillac). A large cabin necessarily has a large frontal surface, and the amount of frontal surface which must be pushed through the air is one of the major obstacles to achieving high speed with a reasonable amount of horsepower. Streamlining helps, but mass is mass, and it takes a great deal of thrust, requiring a lot of horsepower, to move a large container through the air at high speed.

In addition to the cabin, the fuselage holds the engine and its systems in the front while the tail cone supports the tail assembly. Hidden from view under the cabin floor and elsewhere are all manner of cables, pulley and pushrods which connect the levers, knobs and handles used by the pilot to control the airplane with the control surfaces.

Fuselage construction falls into two basic categories: monocoque, and steel tube plus covering. Monocoque refers to the kind of construction in which the aluminum skin of the fuselage comprises the load-bearing structure, reinforced by bulkheads and stringers. The steel-cage type, usually referred to as *truss*, is a combination of a cage built of steel rods and covered with aluminum sheet, wood or fabric. And then there are any number of combinations of the two, usually referred to as semi-monocoque. From the pilot's point of view the type of construction is of no particular consequence, though the combination of steel-truss and sheet aluminum, such as is used in the Mooney aircraft, is probably the safest in the event of an accident, while wood and fabric construction involves considerable maintenance problems which are not applicable to all-metal aircraft.

The Wings

The *wings* are the most important and, in many ways, the most critical component of an airplane. They are, after all, what makes it fly. In simple terms, the wing is an *inclined plane that rides on a cushion of air*. This is totally unscientific and not actually quite true, but the real explanation of how a wing works with respect to the air through which it flies involves an understanding of *Bernoulli's Theorem*, the theory of circulation, and a lot of related scientific gobbledygook which only serves to confuse the issue. From the pilot's viewpoints, it is more sensible to think of the wing as something which tries to push the air down and which, in turn, is held aloft by the refusal of the air to be compressed below it. This effect is most easily understood if we think of air as if it were water because we can see water but we can't see air.

A sheet of metal, being heavier than water, when placed into a body of water will simply sink to the bottom. Similarly, an aircraft wing will fall to the ground if there is no movement. But take that same piece of metal and pull it through the water at high speed, and you'll see that it will float in an attitude of an inclined plane, with the front being slightly higher than the rear. (This principle makes possible the great sport of water skiing.) The same thing happens with a wing. When it is moved through the air at a given speed, it will remain airborne because the flow of the air produces lift.

Wings come in all kinds of shapes and cross sections. Some are rectangular, others are tapered. Some have more curvature at the top than at the bottom, others are of a symmetrical cross section. Some are referred to as *laminar-flow* wings, and in recent years there has been a lot of talk about *supercritical* wings. Unless you're a frustrated airplane designer, forget the whole thing. As pilots we're stuck with whatever type of wing the manufacturer has hung on the airplane we're flying, and there is absolutely nothing to be gained by worrying about whether, in theory at least, another type of wing might actually result in improved performance. As long as it produces lift and doesn't fall off in turbulence, that's really all we have to worry about.

And then there is the placement of the wings. There are low-wing airplanes and high-wing airplanes. Again, it really doesn't seem to make any difference. Low-wing airplanes are generally considered to look sexier, and, as a general rule, high-wing airplanes appear to be a bit more forgiving of sloppy pilot technique. At least I have found this to be true of the various high-wing Cessnas, and with Cessna being the principal company still manufacturing high-wing light aircraft, there is little other means of comparison. (The high-wing Tripacer, manufactured by Piper years ago, was far from a forgiving airplane.)

In practice, the primary difference between the two types of wing arrangement is *pilot visibility*. The wing, by necessity, is an obstruction to vision. In high-wing airplanes (Fig. 3-1) you can't see what's going on above you to the right and left, and in low-wing airplanes (Fig. 3-2) you can't see what's going on below unless you throw the airplane into a very steep bank. On the whole, I prefer the visibility in low-wing aircraft, though the no-longer-in-production Cessna Cardinal (Fig. 3-3) had excellent pilot visibility because its high wings were set farther back than those of the other high-wing Cessnas.

And visibility, once considered relatively unimportant, is getting increasingly critical with the proliferation of traffic in and near airport areas. But *all* aircraft have blind spots, and it is up to the pilot to figure out a way to see around them.

Fig. 3-1. A typical high-wing aircraft (courtesy of Cessna).

In addition to the job of creating lift and thus causing the airplane to fly, the wings are used for one other important purpose: carrying fuel. In virtually all modern airplanes, the fuel tanks are in the inner (and sometimes the inner *and* outer) panels of the wings. These tanks may be rubber or rubberized metal containers set into the wings, or the wings themselves may have been leakproofed, being then used as so-called integral fuel tanks, or "wet wings."

The Stabilizers

Attached to the tailcone are the *vertical* and *horizontal stabilizers* (except when the latter are *stabilators*, which will be discussed later). Both are airfoils and their primary purpose is to keep the tail of the airplane behind its nose. This may sound funny because an airplane, after all, cannot be bent. But without the stabilizers, the entire contraption would be inherently unstable (therefore the name) and would tend to wobble around in the air. The primary purpose of the vertical stabilizer is to overcome the tendency of the airplane to *yaw*, meaning to turn either right or left without banking. And the horizontal stabilizer performs that same function with reference to *pitch*, meaning the tendency of the nose to either drop or climb.

Some vertical stabilizers are further assisted in their task by so-called dorsal fins, lengthwise protrusions attached to the top (and in some cases the bottom) of the tail cone. Their purpose, too, is to act upon the slipstream in a manner which will help maintain directional stability.

In recent years it has become fashionable to equip light aircraft with so-called T-tails. A *T-tail* is exactly the same as a

Fig. 3-2. A typical high-performance low-wing aircraft (courtesy of Beech).

Fig. 3-3. The Cessna Cardinal with its set-back wing and greatly improved visibility (courtesy of Cessna).

conventional tail except that the horizontal stabilizer (or stabilator) is attached to the top of the vertical stabilizer instead of to either side of the end of the tail cone (Fig. 3-4). This arrangement does result in a slight difference in the handling quality of the aircraft for two reasons. One is that the horizontal stabilizer is considerably less affected by the propeller blast, making it somewhat less effective during power-on slow flight. The other is experienced primarily during landing. When an airplane is quite close to the ground it flies in what is known as *ground effect*. The air between the horizontal surfaces and the ground is compressed, thus producing an increase in the generation of lift. In aircraft with conventional tails, this effect is more or less equal on the wings and the horizontal stabilizers. But when the latter is moved up to the top of the vertical fin, it remains clear of the ground effect and then much greater control movements (back pressure) are needed in order to keep the tail low and the nose high.

One model family among the popular light aircraft is equipped with what is referred to as an all-moving or *all-flying* tail. That is the Mooney (all M-20 models from the Mark 20 through the Turbo 231). Here the horizontal and vertical stabilizers are attached to a movable rear portion of the tailcone which swivels around a horizontal axis in response to a push or pull on the control wheel, thus, in pitch, performing the combined tasks of the horizontal stabilizer and the elevator.

The Ailerons

And that wraps up our discussion of the fixed components of the airframe. Let's look now at the movable components.

All movable components of an airplane (with the exception of certain types of leading-edge slats) are designed to be operated by the pilot for the purpose of controlling direction, altitude and speed of the aircraft. Again, a thorough understanding of exactly how they work will help to make one a better pilot.

The *ailerons* are the primary control surfaces of an airplane. They are located at the outer portion of the rear of the wing and, when moved up or down, cause the airplane in flight to bank. A turn to the right of the control wheel will cause the right aileron to be raised and the left one to be lowered, and vice versa. The upward deflection of the aileron causes the wing to drop because, a) the slip stream across the top of the wing pushes against it, and b) the more important and technically more correct reason is that it causes a loss of the ability of the wing to generate lift. Still, for the layman

Fig. 3-4. On T-tail aircraft the horizontal stabilizer is attached to the top of the vertical stabilizer (courtesy of Beech).

(and the average pilot is a layman when it comes to aerodynamic theory and law) it is easier to visualize that air pressure on the top surface of the aileron pushes the wing down.

On the other wing the exact opposite is taking place. Here, the downward deflection of the aileron actually *increases* the lift-generating capacity of the wing, or, in simpler terms, the air striking the underside of the aileron pushes the wing up.

But why do we want to bank in the first place? Here we come to what is probably one of the two most important laws of aerodynamics. To put it quite simply, an airplane cannot turn, in other words, change direction horizontally, unless it is in a bank. As long as the wings remain level the airplane will want to continue to fly in the direction in which it has been flying, even if, through incorrect use of the rudder, the nose of the airplane is forced to either the right or left. In the chapters on flying technique we'll examine the interrelation of ailerons and rudder at some length. For the time being suffice it to say that an airplane must bank in order to turn, and the ailerons are used in order to achieve such a bank.

(Some airplanes are equipped with something known as *drooped* ailerons. In these installations, both ailerons can be raised or lowered simultaneously without losing the ability to use them for directional control. Such a slight raising of the ailerons will tend to marginally increase the maximum cruising speed, while a slight lowering results in a slight increase in the amount of lift being generated, thus improving slow-flight characteristics.)

The Rudder

The *rudder* is one of the most misunderstood controls on an airplane. This misunderstanding probably results from the fact that we like to equate the rudder on an airplane to that on a ship. On a ship, in order to initiate a turn, the rudder is deflected, causing the ship to turn in the direction of that deflection. But here, equating the action of the air with that of water gets us into trouble. A rudder deflection alone does *not* cause the airplane to initiate a turn. What it does do is to yaw the tail to one side and the nose to the other, while the airplane, somewhat awkwardly, tries to continue flying straight ahead. This is called *skidding*. What makes understanding the funtion of the rudder so difficult is that it is *used* during the turn, but it does not *cause* the turn. Visualizing its function would probably be easier if it wasn't called *rudder* in the first place. It is, in fact, not a rudder at all, but rather a means of increasing the effectiveness of the vertical stabilizer at times when the effect of that fixed surface is insufficient to eliminate a tendency of the airplane to yaw due to a variety of causes which will be discussed later. In fact, an airplane could fly and be controlled, somewhat less efficiently, without a rudder. One model, the Ercoupe, was initially built without a movable rudder and did quite well.

The Elevators

Here we have another control that has been misnamed. The *elevator* is not an elevator at all. It is, in fact, the primary means of controlling the speed with which the airplane flies through the air. Its right name should really be airspeed control.

You may think that this is a bunch of nonsense. Obviously, if you pull on the control wheel, deflecting the elevator surface upward, the nose of the airplane will be forced upward, causing the airplane to climb. Conversely, pushing on the control wheel and thus forcing the elevator surface downward will cause the nose to drop, resulting in descent. This is both true and false.

Pulling on the control wheel will raise the nose and, for a brief moment, will cause the airplane to climb. But even though we hold the wheel back and continue to do so, the airplane will not *continue* to climb. What it *will* do is slow down and then continue on in level flight, mushing through the air in a nose-high attitude. As a matter of fact, if we pull back far enough, it will start to descend despite the fact that the nose may be pointing high above the horizon.

When we push forward on the wheel, the opposite occurs. Granted, at first it will lose a few feet of altitude, but while doing so it will speed up and as the increase in speed produces added lift, it will attempt to return to level flight. In this case, though, if we keep pushing hard on the wheel, we can force the airplane into a continuing dive which rapidly will build up speed to exceed the structural capability of the airplane because here we have the gravity pull of the earth working *with* us rather than *against* us.

If the elevator is, in fact, the airspeed control, what, then, is the throttle? The throttle controls the amount of thrust produced by the engine through the action of the propeller. It does not necessarily result in speed as is easily proven by the fact that an airplane can be stalled with the engine and the propeller producing full power. Pilots, especially new student pilots, always find it difficult to think in terms of the elevator as an airspeed control, and many an unplanned collision with the terrain has been the result of instinctively misusing the elevators to gain altitude rather than speed.

If the elevator does not cause an airplane to either climb or descend, what does? The throttle does! Prove it to yourself by this simple experiment. With the airplane trimmed for cruise and the engine operating at, say, 55 percent of power, shove the throttle all the way in without any pressure on the control wheel whatever. Within seconds the airplane will start to climb and it will continue to climb until the reduced density of the air at higher altitudes causes a reduction in engine power output. Conversely, again trimmed for straight and level flight, reduce the throttle and the airplane will start to descend. In this instance it will drop its nose as some airspeed is lost, then speed up until the original airspeed is regained, and then will raise its nose again somewhat and will continue at that original speed in what is probably a very shallow descent unless the throttle setting is further reduced.

This, then, brings us to the next subject.

The Trim Tabs

Let's first clarify what is meant by *trimming* the airplane. The term refers to the ability of the pilot to adjust the elevators (and in some airplanes the rudder) to a position which allows straight and level flight or a continuous climb or descend at a given rate without the need to make any corrections with the control wheel (or the rudder).

There are a number of ways in which this is accomplished. The most common is a so-called moveable *trim tab*, a small surface on the trailing edge of the elevators (and rudder) which can be moved up or down, causing a relatively larger deflection of the elevators (or rudder). A wheel or handle in the cockpit controls the position of these trim tabs and must be operated by the pilot. In order to trim the elevator up resulting in a raising of the nose, the actual trim tab is moved down. Then, the airstream hitting the trim tab will cause the elevators to rise, and their surface, being vastly greater than that of the trim tab, produces the nose-up reaction. (Rudder trim works the same way, only in the vertical plane.)

Now, accepting the fact that the elevator is the speed control, it logically follows that the trim can be used to cause the airplane to fly at any desired speed down to very slow flight (though obviously above stall speed). Most pilots learn early during their flight training to trim their airplanes for level cruise because so much time is spent in cruise and it would be annoying to have to hold the control wheel in a given position in order to maintain altitude. What many instructors fail to teach is trimming the airplane during climb and descent. During climb it is bad enough that we have to continuously hold right rudder to overcome the so-called torque effect of the engine running at full power (unless, of course, the airplane is equipped with rudder trim) but also having to control wheel back in order to maintain a steady rate of climb is a nuisance. Therefore, trimming the nose up will relieve the pilot of that problem.

You may now feel that I have just contradicted myself. If the elevator is the airspeed control, then why do we need to use it to effect continued climb? Well, looking at it realistically, we are not actually using the elevator to effect climb. We are using it to keep the airspeed lower than it would be with the same amount of thrust in level flight. But all that thrust has got to go somewhere, so, since we are purposely using the elevators to keep the speed from increasing, the alternative result is that the excess thrust is converted into an increase in altitude.

Using the trim during the approach to landing is equally important. Planning and eventually executing a good landing involves enough workload for the pilot, and he should not be saddled with the extra chore of continued back pressure on the control wheel to keep the airplane at the right speed on the desired glidepath. If he learns to trim the airplane for slow flight on the downwind leg, it then is an easy matter to gradually increase the nose-up trim during base and on final to set up the desired rate of descent, entirely (or nearly entirely) eliminating the need to put any back pressure on the control wheel. In addition, when it comes time for the flare, the amount of back pressure then needed is much less than it would be if the airplane were still trimmed for level cruise.

In airplanes equipped with stabilators or, in the case of Mooneys, with an all-moving tail, the mechanics of effecting the trimming may be slightly different, but the effect, as far as the pilot is concerned, is exactly the same.

Stabilators

Stabilator is a word coined to stand for something which is a combination of horizontal stabilizer and elevator. In practical terms, what it means is that instead of having a movable surface, the elevator, attached to a fixed surface, the horizontal stabilizer, the whole kit and kaboodle is one surface which can be horizontally rotated around a lateral axis. While highly sensitive pilots might find a difference in control response, the system really works just like the conventional one and it requires little time or genius to get used to it.

But apparently it does take a bit of getting used to, as was proven by a number of mishaps involving the early Cessna Cardinal (Fig. 3-5). Looking and flying most of the time like the venerable Cessna 172 Skyhawk (Fig. 3-6), it did have a fairly large stabilator which apparently was a great deal more sensitive than the stabilizer-elevator combination on the Skyhawk. Cessna pilots, expecting the Cardinal to "fly like a Cessna," would tend to overcontrol during the flare just prior to landing, ending up with the nose on the concrete, a bent prop and a busted nosewheel. Subsequently, Cessna punched some holes into the stabilator to make it less effective, in other words, to make it fly more "like a Cessna." (I flew one of the early models without the holes, and I thought it was a delightful airplane. But apparently not enough people agreed with me, because the Cardinal, sadly, is no longer in production.)

The ability to trim an airplane is one of those subjects which only too often seems to be ignored during flight training. It is, in fact, of such vital importance that it would be difficult to overemphasize it.

The Flaps

Flaps are panels of various sizes which can be extended rearward and downward out of the trailing edges of the inner portions of the wing. In simple terms, their purpose is first to

Fig. 3-5. The original Cardinal had highly sensitive stabilator controls which resulted in flight characteristics which were different than those of other Cessnas and which got some pilots into trouble (courtesy of Cessna).

Fig. 3-6. The Cessna Skyhawk (courtesy of Cessna).

increase the lift-generating capability of the wing, and then, when fully deflected, to create drag. Flaps can be partially or fully deployed, depending on the desired result. Most aircraft manufacturers recommend the use of partial flaps during takeoff in order to increase lift at relatively low speeds and thus to reduce the takeoff run by causing the aircraft to become airborne earlier. Similarly, they recommend the use of partial flaps during the approach to landing for the same reason and possibly full flaps during the final phase of landing to create drag which reduces speed without markedly reducing lift.

Under normal circumstances, the use of flaps in light aircraft is more or less standard procedure and cause for little thought or apprehension by the pilot. But there are times when, for the more experienced pilot, the correct use of flaps can simplify a marginal takeoff or an equally marginal landing.

Let's take the landing first. Assume we've made a perfectly decent approach but, because we're a little too fast, just after flaring a foot or so above the runway, the airplane simply refuses to settle. Here a quick-fix answer is to get rid of the flaps. Retracting the flaps reduces lift and the airplane will sit down firmly on the runway. This procedure is especially effective in excessively clean airplanes in which even a mile or two or extra airspeed is hard to bleed off.

Then there is the time when we have to land in a strong crosswind. Here, flaps tend to provide the crosswind with added surfaces on which to exert its unpleasant power and, in extreme conditions (especially in airplanes with large flaps, such as the single-engine Cessnas), it can become virtually impossible to get the darn thing on the ground, mainwheels first. Don't pay any attention to instructors who like to say that flaps should *always* be used on landing. Baloney! In crosswind conditions, use little or preferably *no* flaps. It takes a bit more flying expertise to get the airplane slowed down to touchdown speed, but without such expertise you shouldn't be attempting a crosswind landing in the first place.

During takeoff, the same holds true if there is a strong crosswind. Forget about the flaps and accept a longer takeoff run in order to achieve the greater speed necessary to permit liftoff without flaps. This may not work on a very short runway, but on such a short runway you may not be able to take off in a strong crosswind in the first place.

A different and a bit tricky use of flaps can occasionally come in handy if, after liftoff, we find that our rate of climb is just a bit too skimpy to clear that fence or bush ahead. Just before—and I mean *just* before—we get to that obstacle, shove down full flaps, *fast*. The airplane will virtually jump up five to ten feet, depending on type. It will clear the obstacle, but it will probably settle back down again afterwards, and it will require a very slow and careful retraction of the flaps in order to exchange a gradual reduction in the lift generated by the wing-flap combination for an equally gradual increase in speed.

Caution: Don't get caught at very slow speed with full flaps in something like a Cessna 150 (Fig. 3-7). You may find that you're what is known as "behind the power curve." What this means is that the engine is incapable of producing sufficient thrust to increase speed with all the drag resulting from those full flaps. The best it can do is to permit the airplane to continue in level flight at a speed which is only a mile or two above stall. Conversely, at this low speed, and being close to the ground, you simply can't afford to try to retract the flaps, because doing so would, at least briefly, result in a loss of altitude, and you simply have no altitude to spare. For all practical purposes you're stuck. You can't speed up and therefore you can't climb. There are really only two ways out: either head toward lower terrain which would give you an opportunity to accept the loss of altitude associated with *very slowly* retracting the flaps and thus effect a gradual increase in speed and, in turn, climbing capability. Or head toward the nearest landing strip which is at an elevation which will permit a landing without having to climb.

Spoilers

Spoilers are what the name implies: Means of disrupting, spoiling, the lift-generating capability of the wing. There are two types of spoilers. One type is normally deployed simultaneously on both wings in order to act as a speed brake. This kind is found on most if not all gliders and sailplanes, which need to be able to make *every* landing at the first try because they have no means of going around. They are also found on high-performance corporate jets and jet airliners in order to provide the drag necessary to reduce the length of the landing run.

The other type is the differential spoiler which on some aircraft takes the place of ailerons. In this system the spoiler on one or the other wing is raised in order to reduce the lift of that wing and, in turn have it drop, initiating a bank. While there is a slight difference in the control response of a spoiler-equipped aircraft in

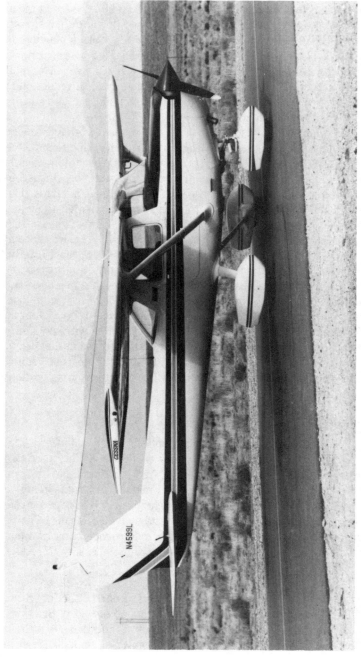

Fig. 3-7. The Cessna 150 has huge flaps which, at slow flight, may result in the pilot finding himself "behind the power curve" (courtesy of Cessna).

contrast to one equipped with ailerons, the difference is not sufficient to be of consequence.

Today, the only commercially produced general-aviation aircraft which uses spoilers instead of ailerons is the Mitsubishi MU-2 turboprop. But NASA has been experimenting with light single-engine and twin aircraft using spoilers, and it wouldn't surprise me in the least if we'd see spoilers on some of the more popular general-aviation light aircraft in the not-too-distant future.

The advantage of spoilers is that they permit the use of the entire length of the trailing edge of the wings to be equipped with flaps, so-called full-span flaps. While at first glance this may seem silly, it is not. All aircraft designers strive for airplanes which will fly as fast as humanly possible with a given amount of horsepower, but at the same time these same airplanes must also be capable of being slowed down to an acceptably slow approach and touchdown speed without stalling or otherwise becoming uncontrollable. Equipping wings with full-span flaps permits a more efficient wing design in the high-speed envelope, while retaining, due to the large flap surfaces, the ability to continue to be controllable at quite low speeds. There are still a number of problems with which NASA is wrestling at the moment, but it is safe to assume that they will sooner or later be solved.

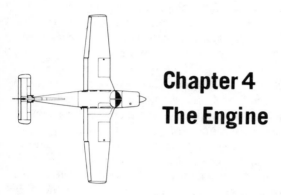

Chapter 4
The Engine

So far we have talked about what is generally referred to as the *airframe*. But every powered aircraft needs an engine, and it would appear worthwhile to spend a little time looking at the various types of power plants which are utilized for the purpose of developing the thrust needed to push the airframe through the air.

Reciprocating Engines

The oldest and most common type of powerplant is the normally aspirated reciprocating gasoline engine, similar, though not identical, to the one used to power your automobile. The term *normally aspirated* means that the engine is not turbocharged (or supercharged). The term *reciprocating* refers to the fact that the up-and-down or back-and-forth movement of the pistons in the cylinders is translated into the rotary motion needed in order to drive the propeller. The engine component which performs this translation from lateral to circular motion is the crankshaft.

Since this book is not intended to turn you into an experienced engine mechanic, let's break down the operation of a reciprocating engine into simple, actually somewhat oversimplified terms.

Inside the engine are a number of cylinders (usually either four or six) and inside those cylinders are pistons. A mixture of fuel and air is fed into each cylinder through a carburetor or injection system, and, when ignited by a spark plug, explodes (or more accurately, burns very rapidly), causing the air to expand, pushing the piston to the far end of the cylinder. On the return trip, the piston expels what is left of that explosion and then, after once more returning to the far end of the cylinder and coming back again, another explosion starts the cycle all over again (Fig. 4-1). This back-and-forth movement of the pistons acts on the crankshaft through a fairly complicated system of levers, turning the crankshaft at a speed relative to the number of round trips made by the pistons.

In most installations the propeller is attached directly to the crankshaft, while in some a gearbox is used to cause the propeller to turn at a somewhat slower rate than the crankshaft.

So far everything I have said equally applies to your automobile engine, except that the crankshaft, instead of turning a propeller, turns the wheels. But there are some differences. First of all, because weight is an important consideration in aviation, the overall weight of aircraft engines is usually only a fraction of that of automotive engines of comparable horsepower. Second, while in your automobile the spark plug is ignited by current derived from the battery (which is constantly being recharged by the generator or alternator), the task of igniting the spark in an airplane is accomplished by a *magneto*, actually a small electric generator driven by the engine and designed specifically for this purpose. Thus an airplane can fly indefinitely without a battery (or with a dead battery), while an automobile will stop running if the generator (alternator) belt breaks, as soon as the battery is drained, which doesn't take very long. Third, while automotive engines are equipped with one spark plug per cylinder, aircraft engines use *two* for each. This redunancy is strictly a safety measure. Thus, if one spark plug gets fouled, the other one will continue to fire, preventing an engine-out situation.

Reciprocating engines must be fed a certain proportion of fuel to air in order to function. The ideal fuel-to-air ratio is approximately 15 pounds of air to each one pound of fuel. But the weight of the air varies at different altitudes and different temperatures. The higher the altitude or the warmer the air, the lighter it is. Thus, a given throttle setting will permit less and less fuel to be burned the higher we fly (or the warmer the air is), causing a reduction in the available power and the resulting thrust. As a matter of fact, most normally aspirated reciprocating engines can produce full power only up to an altitude somewhere between 6,000 and 8,000 feet, assuming standard temperature (which at those altitudes is +3 or −1 degrees C.). Above those altitudes, full power will produce an increasingly diminishing percentage of power. (For a detailed discussion on the effects of altitude and temperature on engine and aircraft performance, see Chapter 11 dealing with density altitude and fuel consumption.)

Turbocharging

The only way to overcome this problem is to find a way to increase the amount of air forced into the engine beyond the amount which is provided by the movement of the aircraft through the air. The answer is *turbocharging*.

Turbocharging, stated simply, is a mechanical means of sucking in a lot of air, compressing it, and then feeding that compressed air along with the appropriate quantity of fuel to the engine. The means of accomplishing this is a small turbine which is driven by the exhaust gases of the engine. It pulls in outside air in huge amounts, compresses it and then forces it into the intake manifold. Because this turbine is driven by energy which is normally simply dumped overboard, it does not use any noticeable amount of engine power.

The result of turbocharging is that the engine can produce full power up to about three times the altitude at which a normally aspirated engine starts to lose manifold pressure. In other words, it negates the detrimental effects of altitude and high temperatures.

Propellers

All reciprocating engines, whether normally aspirated or turbocharged, drive a propeller which produces the necessary thrust. Propeller systems, again, fall into two basic categories: *fixed pitch* and *constant speed*.

A fixed-pitch propeller, almost without exception a two-bladed propeller, has been designed in terms of blade angle to be a compromise between the best angles for takeoff and climb and the most ideal angle for cruise (Fig. 4-2). Being a compromise, it is not perfect during either of these flight profiles, but it has the advantage of being easy to operate and is therefore found primarily on training aircraft and on other so-called low-performance singles. With a fixed-pitch propeller, advancing the throttle simply increases the rpm up to a permissible maximum, and the percentage of power is directly related to the number of revolutions per minute. These propellers are fairly satisfactory at the lower altitudes, but become increasingly deficient the higher we fly.

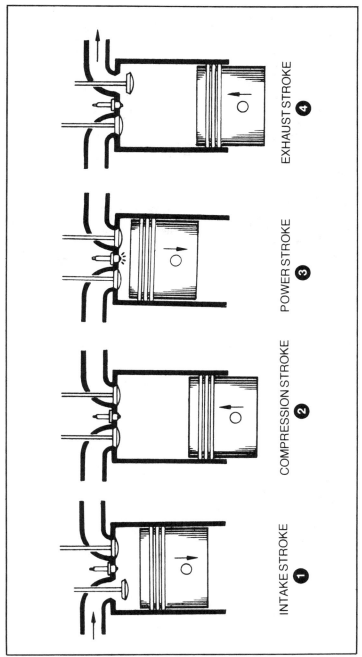

INTAKE STROKE ❶   COMPRESSION STROKE ❷   POWER STROKE ❸   EXHAUST STROKE ❹

Fig. 4-1. The four strokes of a four-stroke reciprocating engine.

Fig. 4-2. Light aircraft with fixed-pitch propeller (courtesy of Cessna).

(Still, I once coaxed a Cessna 175, equipped with a fixed-pitch prop, all the way up to 18,800 feet. So it *can* be done.)

All high-performance singles and all piston twins are equipped with *constant-speed* propellers, using either two or three blades (Fig. 4-3). Constant-speed propellers, also known as *variable-pitch* propellers, are equipped with a governor which controls the turning speed and which can be adjusted by the pilot by means of a control located in the cockpit. What happens in practice is that the blade angle is changed from very slight to fairly extreme, changing the angle of attack of the blades relative to the air through which they move. The effect on performance can be likened to the gearshift in an automobile. Placing the blade angle into flat pitch is comparable to driving in low gear. The engine turns at great speed but the resulting amount of thrust is less than it would be with a greater blade angle, assuming comparable rpm. As a result, the engine doesn't have to work as hard as it would if the blade angle were steeper. Just as low gear is used in an automobile during startup and when climbing steep hills, so the propeller is placed in flat pitch during initial acceleration and the first portion of the climbout until a reasonable amount of speed has developed.

Then, just as an automobile uses the third or fourth gear position in level drive, a gear position in which the engine turns over fairly slowly while the wheels go fast, during level cruise the blade angle can be steepened which results in a reduction in the rpm but without a reduction in the airspeed, because at the greater blade angle the propeller takes larger bites out of the air with each revolution. This not only increases the efficiency of the engine-propeller combination, it also helps to save fuel; with a given amount of fuel being burned during each engine rpm, a reduction in the number of rpm will automatically also reduce the fuel consumption.

Constant-speed propeller installations do increase the workload of the pilot to a certain extent. Instead of having just the one throttle which controls everything in aircraft equipped with fixed-pitch propellers, there is a throttle for the purpose of controlling manifold pressure and another to control propeller pitch.

Manifolds

At this point it might be appropriate to spend a few moments on the subject of manifolds. Technically, a *manifold* is simply a tube

Fig. 4-3. A light aircraft with a three-bladed constant-speed propeller (courtesy of Piper).

49

with one inlet and two or more outlets or with one outlet and two or more inlets. Reciprocating engines have two manifolds, an *intake* manifold and an *exhaust* manifold. The intake manifold takes in air from the outside and distributes it, via the carburetor or the injection system, to the cylinders in order to provide the fuel-air mixture which is appropriate for ignition. The exhaust manifold takes exhaust gases from each cylinder and directs them through the exhaust pipe to the outside.

When we're talking about *manifold pressure*, we are referring to the intake manifold and to the pressure of the air inside that manifold which is available for engine operation. Obviously, in normally aspirated piston engines the manifold pressure cannot be greater than the atmospheric pressure wherever the airplane happens to be at a given moment. (During flight it is, in fact, a *tiny* bit greater than the ambient pressure because of the speed with which the air intake moves through the air. But this increase is too small to warrant serious consideration.) The manifold pressure is measured and expressed in the same terms as is atmospheric pressure, namely in inches of mercury. Now you may ask with a degree of justification, what does mercury have to do with the amount of air? Well, inches of mercury, commonly described as *in hg*, is one of those measurements which were devised years ago and which we have been stuck with ever since. It refers to the fact that a column of mercury in a glass tube will rise a given distance in response to atmospheric pressure exerted on a bowl at the bottom of that glass tube (something like a thermometer). Thus, when this pressure is sufficient to raise that column of mercury 29.92 inches, we have what is known as *standard atmospheric pressure*.

Now back to the engine. As we have discussed earlier, the amount of air available to the engine determines the amount of fuel that can be burned. Thus, operating at high manifold pressure means that greater amounts of fuel are burned, producing greater engine power. As the amount of manifold pressure is reduced, the amount of fuel being burned is reduced, and, in turn, the engine develops less power.

But the amount of power developed by the engine has no direct effect on the propeller rpm in installations equipped with constant speed propellers (except when the manifold pressure is reduced to a point at which there is not sufficient engine power left to turn the propeller at the maximum speed permitted by the setting of the governor). Theoretically, we can operate at very low manifold pressures and very high rpm, or we can do the opposite, raise the manifold pressure up high but keep the propeller rpm at a minimum. In practice this is possible only to a limited extent because the relationship between manifold pressure and rpm has an effect on the health and life of the engine.

Here, again, the comparison with an automobile comes in handy. When an automobile is driven up a steep hill in high gear, the engine must work so hard that it is likely to develop a condition known as pinging. It is the result of something referred to as preignition, a condition, resulting from excessive engine overheating, when the fuel-air mixture ignites before the spark plugs have had an opportunity to fire. In an automobile, this situation can be heard and therefore corrected before serious engine damage results. Not so in an airplane. In an airplane neither preignition nor detonation, a similar but slightly different condition, can be heard by the pilot who is, therefore, unable to take corrective action, and serious engine damage can result.

An airplane operating at high manifold pressure and low rpm is like a car driving uphill in high; it must work excessively hard to drive that propeller with its blades in steep pitch, and if the allowable limits are exceeded, engine damage is sure to result. For this reason, all aircraft manufacturers provide data in the Pilot's Operating Handbook tables which show the various permissible manifold pressure-rpm combinations at different altitudes, also showing the percentage of power produced by these combinations, the amount of fuel burned in terms of either gallons or pounds per hour (gph or pph), and the true airspeed which can be expected to result.

It might be pointed out here that the effect of the number of rpm on the fuel flow is greater than the effect of differences in manifold pressure. Thus, flying at the lowest permissible rpm is always cheaper than using higher rpm and lower manifold pressure to produce the same speed. Using low rpm in cruise also increases the time between major overhauls (TBO. For more about TBOs, see the end of this chapter). The TBO recommended by the manufacturer, though expressed in hours of operation, is actually based on total rpm which can be expected to take place over a given period of time. This figure is usually based on a relatively high rpm figure. Thus, if an engine has a 2,000-hour TBO and that figure is based on 2,400 rpm, the total number of rpm between overhauls is 288 million. Now, if that same engine is habitually flown at, say, 2,000 rpm, it will take 2,400 hours to accumulate that same number of rpm, an increase of 20 percent.

There are three other types of engines used in aircraft. None of these are of any direct interest to most of those who can be expected to read this book. Still, a brief look at each may be called for.

Turboprops

The next step up from the reciprocating engine is the so-called *turboprop* or *propjet* (or in the case of helicopters, turboshaft) engine (Fig. 4-4). It is, in fact, a jet engine which is designed to drive a shaft to which a propeller is attached. (In helicopters, this shaft, through a complicated combination of linkages and gears, drives the main and tail rotors.) Reduced to the simplest of terms, it is a turbine, primarily a single axis, one end of which is designed to be driven by the burning of a mixture of air and jet fuel, while the other end turns the propeller. While the fuel consumption of these engines is vastly greater than that of reciprocating engines, the primary advantage, in addition to greater thrust and, in turn, greater airspeed development, is the extreme reliability. With, for all practical purposes, only one moving part (there are lots of gear boxes and other accessories, but technically they are not an integral part of the engine) there is little that can go wrong or wear out, and, as a result, the TBOs of these engines are two or more times those of reciprocating engines.

The drawback of turboprops in comparison with other jet engines is a degree of altitude limitation. Because propellers lose efficiency in the extremely thin air at very high altitudes, most turboprops operate best at altitudes between 16,000 and 30,000 feet, which puts them above some but not all of the weather and turbulence.

Turbojets

Then there is the so-called *turbojet* or pure jet (Fig. 4-5). In these engines intake air is expanded by the burning of fuel to a point that causes the air to escape out of the rear of the engine at a tremendous speed, thus simply producing thrust as a reaction to that jet blast. (It isn't quite that simple, but for our purposes this explanation should be sufficient.) These engines burn the most fuel of all, and are extremely noisy. As a result they are used in fewer and fewer aircraft (other than military). But they do have an advantage. They do not lose efficiency at altitude. As a matter of fact, the higher such an airplane flies, the better these engines function and the less fuel they burn.

Fig. 4-4. A turboprop aircraft, the Swearingen Merlin.

Fanjets

Today nearly all civilian jet aircraft are powered by what are referred to as *fanjets*. In the simplest of terms, a fanjet is a combination of the principle of the turboprop and that of the pure jet (Fig. 4-6). The fanjet contains a turbine which drives a small, multibladed propeller (fan) which is fully contained within the engine housing and thus cannot be seen unless we look into the engine. The burning of fuel accomplishes two tasks simultaneously. It drives the turbine and in turn the fan at very high speeds. It also produces a certain amount of pure thrust which bypasses the turbine and the fan, which is why some of these engines are referred to as *high-bypass* engines. This pressurized air which develops a certain percentage of the total engine thrust (less than the fan), in the process of bypassing the turbine and fan on all sides, acts as a barrier to sound. Thus these engines are a great deal quieter than the pure jets. They also are much more fuel efficient, an important consideration in these days of high fuel costs. These engines, too, containing, in principle, only one moving part, are extremely reliable, resulting in very high TBOs.

TBO

What's a TBO? The term stands for time between (major) overhauls. It is always expressed in hours, meaning hours of operation. It means that after the engine has been operated for a given number of hours, it should be taken apart, thoroughly checked, worn parts should be replaced, and then the whole thing should be put together again, theoretically being almost like new. A TBO is not a mandatory requirement. It is a suggestion based on past experience, a suggestion made by the manufacturer. Because past experience with a given type of engine is an important consideration in determining TBOs, the hours involved may be increased from time to time as it is experienced that, that particular engine does not appear to develop any troubles within the number of hours of the previously applicable TBO.

A TBO is no guarantee. By recommending a certain TBO the manufacturer does not guarantee that nothing can possibly go wrong with that engine until the TBO limit is reached. Some engines, for any number of reasons (not the least of which may be pilot techinque), will konk out or develop troubles way short of their TBO, while others may continue to operate perfectly for hundreds of hours beyond it.

Fig. 4-5. A turbojet aircraft, the Hawker 600 (courtesy of Beech).

Fig. 4-6. A fanjet aircraft, the Learjet 35 (courtesy of Gates Learjet).

It is up to the pilot to monitor engine performance carefully at all times, and not to simply ignore the engine because the TBO limit is still some hundreds of hours in the future. Modern engines don't simply quit. Whenever some problem is in the process of developing, the watchful pilot can usually detect some abnormality, no matter how minor, in instrument readings or possibly just sound or feel. Whenever there is some suspicion that something is amiss, it's time to have an expert look at the engine. It may one day save the life of the pilot and whoever else is on board.

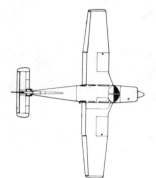

Chapter 5
Air-Data and
Engine Instruments

The average non-pilot, used to the Spartan look of most automotive instrument panels, is likely to take one look at the instrument panel of even a simple training airplane and think that it must take a degree of superhuman perspicacity to understand the purpose of all those dials, knobs and buttons, and to learn how to use them properly (Fig. 5-1). And it is quite true, at first glance the front office of an airplane can be a bit bewildering. But it's really not that complicated, as is proven by the fact that eventually all pilots, even those who are not mental giants—and there are quite a few—learn to deal with them. Cockpit instruments and controls fall into three basic categories: *Air-data* instruments register the attitude, speed, altitude, and related parameters. *Engine* instruments and controls relate to everything that is in one way or another connected with engine performance. And the *electronics* used for communication and navigation. In addition, there are convenience items such as heat, fresh-air controls, and so on.

In this chapter we'll look at all of these except the communication and navigation systems which are the subject of a separate chapter.

Airspeed Indicator

The airspeed indicator (ASI) should be considered the most important of all the air-data instruments (though not all flight instructors would agree with this), because as long as airspeed is

Fig. 5-1. A typical instrument panel of a light single-engine aircraft (courtesy of Cessna).

maintained within the allowable limits, it is nearly impossible for a pilot to get into trouble. In principle, an airspeed indicator (Fig. 5-2) measures the difference between the atmospheric pressure of still air against the air pressure caused by the moving air as it enters the pitot-static system through the pitot tube.

The pitot tube is a usually L-shaped protrusion attached to the wing in an area which is clear of the propeller blast. It has a small opening pointed forward. The ram air resulting from the movement of the aircraft enters this opening under pressure and is forced through a series of hollow pipes into the airspeed indicator. But this pressure alone would not produce a meaningful reading if it could not be compared to the static pressure, the atmospheric pressure in an area of undisturbed air. For this purpose aircraft are equipped with so-called static vents, tiny openings in the sidewall of the aircraft where the air is relatively undistrubed. These static vents and their associated piping represent the other half of the pitot-static system.

With information obtained from both these sources the ASI is able to tell us how fast the airplane is moving through the air. Except most of the time it is not telling us the truth. The reading it provides is known as *indicated airspeed* because that's what the needle says. But, depending on altitude, it tends to differ dramatically from *true airspeed*, the speed at which the airplane is actually moving through the air. Indicated airspeed and true airspeed are commonly referred to as IAS and TAS respectively,

Fig. 5-2. Airspeed indicator (courtesy of Intercontinental Dynamics Corp.).

and we'll be using those abbreviations from now on. The reason for the difference between IAS and TAS is that the thinner the air is, the less pressure it exerts on the pitot-static system, thus producing a lower-than-true reading. For instance, a 100-mph reading at 1,000 feet at standard temperature (13°C.) represents a TAS of 101 mph. But that same reading under comparable conditions at 10,000 feet represents a TAS of 116 mph. There are two ways in which to determine the TAS. One, the simplest but least precise, is provided by most of the newer airspeed indicators. They are equipped with a rotatable dial. When this dial is set to the current altitude, the needle, instead of indicating IAS, will show the approximate TAS. A more precise result can be obtained by using the E6b circular aviation computer, or any of the more sophisticated electronic aviation computers which have more recently invaded the market.

It is important to understand the difference between IAS and TAS with reference to the behavior of the airplane. For instance, the speed at which the airplane stalls, or the speed at which it is ready to be lifted off the runway, is always the same IAS, regardless of temperature, altitude or airport elevation, though the TAS will vary considerably with these conditions. It is therefore recommended that the pilot pay attention to the IAS, not the TAS, during maneuvers for which airspeed control is critical. This means that in aircraft which are equipped with airspeed indicators with those rotating dials, he should always return the dial to its sea level reading before takeoff and prior to initiating an approach for landing. If he doesn't, he may be reading TAS while thinking that what he is looking at is IAS, causing him to inadvertantly stall because the reading is higher than it would have been if it had been referenced to sea level.

Altimeter

The altimeter (Fig. 5-3) is installed for the purpose of telling us how high we are above sea level, but it, too, does not always tell the truth. Its principle of operation is similar to that of the airspeed indicator, except that instead of comparing ram air to ambient static pressure, it compares that ambient static pressure with a small amount of air trapped within the instrument. At sea level, the atmospheric pressure and the pressure of the air trapped in the instrument are equal, producing a reading of zero. As the airplane climbs, the atmospheric pressure decreases and, by comparing it

Fig. 5-3. Altimeter (courtesy of Aero Mechanisms).

with the trapped air, the reading increases, showing us the altitude, in feet, at which we are.

The reason why altimeters tend to lie is that the atmospheric pressure is determined not only by altitude but also by the prevailing temperature (and to some unimportant degree by humidity). In order to allow for these variations, the altimeter is equipped with a little window and an adjustable knob. By turning the knob we can set the current atmospheric pressure, adjusted to sea level, into that little window and arrive at the correct altitude reading. The usual way in which this is done is as follows: Prior to takeoff, knowing the published field elevation of the departure airport, the pilot turns the knob until the altimeter needle indicates that known airport elevation. The numbers then appearing in the little rectangular window represent the actual atmospheric pressure at that moment, adjusted to sea level. As long as there is no change in that atmospheric pressure, the altimeter will produce a reliable reading. (If the numbers in the little window do not exactly coincide with the information provided by ground control, the reason is that either the altimeter is not perfectly calibrated, or that the information given by ground control is not current.)

The term "adjusted to sea level" might require a bit of explanation. The standard atmospheric pressure is considered to be 29.92 in hg. But that is true only at sea level. At any altitude above sea level, it is less. For instance, I live in Santa Fe, New Mexico, at an elevation of 7,000 feet. Here the actual standard atmospheric pressure is in the neighborhood of 24.50 in hg. But since this would become incredibly confusing, all barometric-pressure reports are always automatically adjusted to sea level (unless otherwise indicated).

What causes the altimeter to give erroneous readings is the fact that the atmospheric (barometric) pressure en route is not necessarily the same as it was at our departure point. All we can do to counteract that is to listen to the periodic weather reports while en route, and to adjust the altimeter to whatever changes in that pressure are reported. But even this will only provide an approximation, and we'll always be safer by flying a little higher than the minimum altitude needed in order to clear terrain or a given obstacle.

Altimeters are equipped with several hands or needles. The largest one usually shows altitudes from zero to 1,000 feet. The second hand then shows altitudes in thousands of feet. Thus, say, we are at 6,500 feet; the long needle will be sitting on 5 and the shorter one on 6. To indicate altitudes above 10,000 feet, a third, usually still shorter needle appears. It will point to 1 at 10,000 feet, to 2 at 20,000 feet and so on. Some of the more sophisticated altimeters have slightly different types of readout (Fig. 5-4), but the principle is always more or less the same.

Since the two greatest safety factors in aviation are speed and altitude, I personally consider the altimeter the *second* most important instrument in the cockpit. Any pilot must be able to glance at its face and comprehend its meaning at a glance, as inadvertant misinterpretations, such as thinking you are at 11,000 feet when, in fact, you are at 1,100 feet, can be *lethal*. (Never mind the oft-quoted mother who told her pilot son to be sure to fly "low and slow.")

Compass

Now that we have a means of knowing how fast we're going and how *high* we are, the next thing we'll want to know is the

Fig. 5-4. A sophisticated altimeter for high-performance aircraft (courtesy of Aero Mechanisms).

direction in which we're flying. For this purpose the aircraft is equipped with a compass, or rather, *two* compasses. One is a magnetic compass and the other a gyroscopically-driven compass, usually referred to as the *directional gyro* or *DG*. The two work on entirely different principles, but both are needed in order to fly in a given direction for any length of time.

The magnetic compass (Fig. 5-5) consists of a rotating unit suspended on the tip of a needle inside a bowl filled with liquid. When left undisturbed it will always adjust itself to the magnetic north (there is a difference between magnetic north and true north, but more about that later), thus indicating the magnetic heading toward which the nose of the aircraft is pointed. But the secret word here is "undisturbed." The moment we start to either climb or descend, or we decide to bank in one direction or the other, the compass begins to act as if it had consumed an overdose of martinis. It swings wildly this way and that and its reading becomes totally meaningless until the aircraft has returned to steady straight and level flight, and even then it takes it a few moments to get its act together. As a result, using the magnetic compass for anything resembling precision navigation is a hopeless task. Then what is it good for? The answer to this will become apparent as we examine the second compass, the DG.

A DG (Fig. 5-6) is a compass which is entirely unaffected by the magnetic field of the earth. It is simply a rotatable compass card

Fig. 5-5. A magnetic compass mounted in the center of the windshield of a Mooney 201.

Fig. 5-6. A directional gyro (courtesy of AIM).

which can be manually set to any direction, and which will then react to the changes in direction of the aircraft, and do this with considerable precision. It will allow the pilot to make turns to a given new heading and show him exactly when to roll out of the turn in order to take up that new heading. It is quite foolproof, assuming that it has been set correctly in the first place. But it, too, has a quirk. Its peculiarity is known as *precession*. Because there is no such thing as a mechanism which is entirely free of friction, the DG will gradually deviate from the setting to which it had been set originally, and it is up to the pilot to make adjustments from time to time. Here is where the interaction of the two compasses comes in. Every 10 or 15 minutes the pilot should, at a time when he is flying a steady straight-and-level course, compare the readings of the magnetic compass with those of the DG, and if they don't coincide, which they probably won't, he should then adjust the DG to the reading of the magnetic compass.

It should be pointed out here that during changes in direction the two compasses rotate in opposite directions. This is a pecularity which initially tends to disturb and confuse the novice pilot, but one soon gets used to it and eventually one doesn't even notice it any longer.

But a compass simply shows the *heading*, the direction in which the nose of the aircraft is pointing. It doesn't actually tell us where we're going. Now you may say that since an airplane certainly can't fly sideways, if the nose is pointing, say, east, we're certainly heading east. Not so. There are two concepts which influence our direction of flight with reference to the ground. One is known as *heading*, and that is the one which coincides with the way the nose of the airplane is pointing. The other is known as *course*, and that is the direction in which we actually move across the surface of the earth. The difference between the two is wind (Fig. 5-7).

An airplane flies within an envelope of air, and this envelope itself is moving in one direction or another. It therefore follows that if an airplane heading *east* flies within an envelope of air moving *south*, the airplane will be crossing the ground in a *southeasterly* direction. Suffice it to say here that the compass can only tell us our *heading* but not our *course*. We'll delve somewhat deeper into the relationship between heading and course in the chapter on navigation.

Turn-and-Bank and Needle-and-Ball Indicators

The needle-and-ball instrument (Fig. 5-8) was one of the earliest air-data instruments, and in its day it probably served an important purpose. More recently it has usually been replaced by the so-called turn-and-bank indicator (Fig. 5-9) which looks different but serves the same purpose. The needle portion of the needle-and-ball and the upper portion of the turn-and-bank are designed to indicate the degree of bank of the aircraft during a turn. Prior to the time when artificial horizons (see below) became, for all practical purposes, standard equipment on aircraft, these indicators helped pilots to maintain a continuous and unchanging rate of bank, and, as a result, a continuous and unchanging rate of turn. The ball portion, a slightly curved glass tube containing a ball

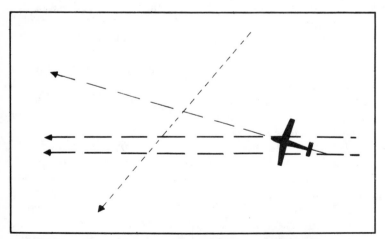

Fig. 5-7. Course is the direction of flight of the aircraft. Heading is the direction in which the nose of the aircraft is pointing. The difference between the two is caused by the wind component.

Fig. 5-8. Needle and ball (courtesy of AIM).

suspended in liquid, is a part of both types of instruments. It simply indicates whether a slip or skid is present during the turn. During a perfectly coordinated turn, one during which the rudder is used to perfection, the ball will remain in the center of the curved tube. If the aircraft starts to skid, indicating that rudder is being used too much, the ball will drift to the end of the tube away from the direction of the turn. In a slip, meaning that insufficient rudder pressure is applied, the ball will move toward the end of the tube which is in the direction of the turn. In each case, using a certain amount of rudder pressure on the side which coincides with the offset of the ball will tend to return the aircraft to a condition of a coordinated turn. In other words, right ball-right rudder; left ball-left rudder.

Today, with virtually every aircraft being equipped with a much more useful artificial horizon, the ball portion of the turn-and-bank (or needle-and-ball) instrument is really the only one which still has any important meaning for the pilot. Don't get me wrong. A momentary deflection of the ball in one direction or

Fig. 5-9. Turn-and-bank indicator (courtesy of AIM).

the other, while possibly being indicative of some sloppy flying, is of no particular importance. On the other hand, if the ball remains off center during a prolonged period of cruise, then either the airplane is out of rig, or, if equipped with a rudder trim, the trim is set incorrectly. Either way, the situation will reduce cruising speed and, in turn, increase the amount of fuel used to cover a given distance.

Artificial Horizon

The artificial horizon (often described as AH) is a gyroscopic instrument which shows the position of the aircraft in terms of bank and pitch with reference to the actual horizon. It is an extremely useful instrument and virtually (through not totally) irreplaceable for flight in instrument conditions when reference to the actual horizon is no longer available. Its indications tend to be a bit confusing to novice pilots and require getting used to.

The artificial horizon consists of a stationary aircraft symbol and an equally stationary dot (Fig. 5-10). These are displayed against a moving background, for all practical purposes simply a moving line, which represents the position of the horizon in relation to those stationary symbols. If the aircraft banks to the right, the moving line will bank to the left, showing that the right wing of the aircraft symbol is closer to that horizon line than is the left wing. If the nose of the aircraft is pulled up, the horizon line will move down, placing the dot above the horizon line. While this sounds perfectly realistic and simple as we talk about it here, in the aircraft it always tends to seem a bit odd at first, and it takes a certain amount of practice to quickly interpret what appear to be contradictory displays on the AH.

Again, it is of vital importance for the pilot to learn to interpret the readings of the AH instantly, because if he should find himself in what is casually being referred to as an unusual attitude, either in IFR conditions, or in conditions of haze or over mountainous terrain, where the actual horizon is either not visible or unreliable, and instant and, more important, correct reaction may be the difference between safety and disaster. Just for practice, study the accompanying illustrations (Figs. 5-11 through 5-13) and figure out the correct corrective action which must be taken in order to return to straight and level flight.

Caution: Quite often a so-called unusual attitude may result in exessive airspeed. Before pulling hard on the control wheel in order to correct for an extreme nose-low condition, take a quick

Fig. 5-10. Artificial horizon.

glance at the airspeed indicator. If the needle is hovering around the red line, it means that a too rapid control movement might overstress the airframe. You'll still have to pull back, but do it slowly and gradually to give the airspeed a chance to bleed off.

Vertical-Speed Indicator

The vertical-speed indicator (VSI) is an instrument which operates on a principle which is somewhat similar to that of the altimeter (Fig. 5-14). It tells the pilot at what rate, in terms of feet per minute (fpm) the aircraft is climbing or descending. In level flight the indicator needle remains horizontal. As a climb is initiated it starts to point upward and the dial is calibrated in hundreds of feet. When the aircraft loses altitude it points downward in a similar fashion.

One of the peculiarities of the VSI which the pilot must remember is that its indications tend to lag a bit behind changes in

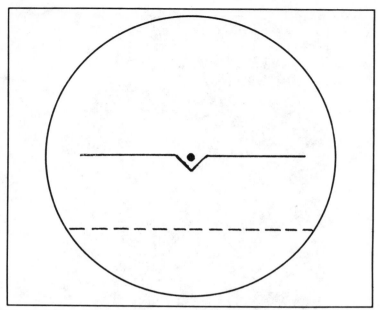

Fig. 5-11. The indication provided by the artificial horizon when the nose of the aircraft is pointing high up into the sky. In this instance the aircraft is likely to be on the edge of a stall.

the rate of upward or downward movement of the aircraft. Therefore, if he is trying to set up a steady rate of either climb or descent, he'll always have to wait a few moments before the instrument reading becomes reliable. (Actually all of the instruments which operate through the pitot-static system exhibit a slight lag, but that of the VSI is greater than that of the others.)

In day-to-day operation the VSI is one of the less important instruments and it is only peripherally included in the pilot's scan of his instruments. Its primary advantage is the ability to assist the pilot in setting up a given rate of climb or descent, and to help him to maintain that rate. The time when this proves most helpful is when we have figured out that a given rate of descent will get us to the pattern altitude at our destination within a certain period of time. It is then that maintaining that specific rate of descent proves most valuable and fuel efficient.

Outside Air Temperature Gauge

The outside-air-temperature gauge (OAT) is simply a thermometer the sensing portion of which is outside the aircraft while its readout is in the cockpit. Aside from simply satisfying the curiosity of the pilot about the temperature outside the aircraft, it provides us with the ability to figure out our TAS with a degree of precision. The computation which transposes IAS to TAS, if the result is to be precise, must, take into account the pressure altitude, the outside-air temperature and the IAS.

The only other time when its reading can become critical is at times when there is a degree of visible moisture in the air and the temperature reading hovers at or a few degrees above freezing. Under these conditions, ice tends to accumulate on the airframe (and, under extreme conditions, on the propeller), and the pilot would do well to seek out either warmer or much colder temperatures by either climbing up to a higher altitude or by dropping down lower.

Clock

The clock which is usually to be found in the extreme left-hand corner of the instrument panel is of minor importance in terms of telling the pilot what time it is. Since in aviation, the fuel on board

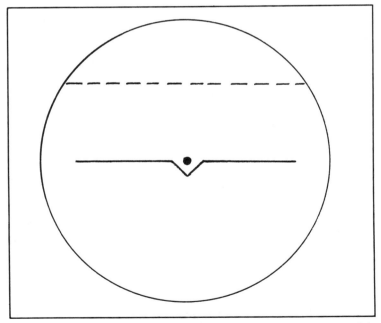

Fig. 5-12. In this indication by the artificial horizon the nose of the aircraft is excessively low, indicating a steep dive.

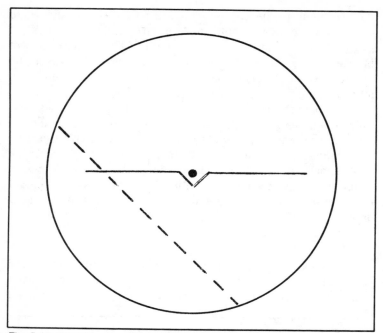

Fig. 5-13. Here the aircraft is in a 45-degree left bank with the nose pulled up high above the horizon.

is usually thought of in terms of hours rather than gallons or pounds, the clock acts as a secondary fuel gauge. As long as we know how many hours of fuel we have on board at the start of the flight, checking the clock from time to time will give us a reading of how long we can remain airborne before the need for a fuel stop becomes critical.

In addition, the IFR pilot needs a clock in order to fly most instrument approaches, many of which depend on flying each leg of the approach for a given number of minutes and seconds at a given airspeed.

The third practical application is in determining the ground speed being made good at any time during the flight. Since TAS has nothing to do with the speed with which we are moving across the surface of the earth, the only reliable way to know our ground speed (other than by using a DME) is to check on the amount of elapsed time between two checkpoints, the distance from one such checkpoint to the other being known to us. The winds aloft tend to be changeable, and winds-aloft forecasts are most of the time less than reliable. It would be foolish, therefore, to assume that, just because the forecast has called for a 15-knot headwind, our ground speed is 15 knots slower than our TAS. It behooves the pilot to check on his ground speed every half hour or so if he wants to avoid unpleasant surprises somewhere along the line.

Stall Warning

The stall-warning device is not technically a cockpit instrument. There is nothing the pilot has to do to operate it. It simply warns him, by means of a blaring sound and/or a red light on the instrument panel, that his aircraft has slowed to within a few miles of the stall. It is operated by a little gimmick in the leading edge of the wing which reacts to changes in the air pressure (the relative wind).

Since the stall warning indication starts *before* the actual stall is reached, it is possible to slow-fly an airplane with the thing blaring into our ears. But it must be remembered that the warning sound (or light) does not change if the speed then diminishes to the actual stall. The only further warning available to the pilot is the feel of the airplane.

Angle-of-Attack Indicator

This instrument, which *should* be in *every* airplane but, in fact, is rarely found in light aircraft, has been discussed at some length

Fig. 5-14. Vertical-speed indicator (courtesy of Smiths).

earlier. It operates on a principle similar to that of the stall warning, but instead of simply warning us of the approaching stall, it provides a cockpit readout showing the angle of attack throughout all flight profiles. By paying attention to the indication of the needle, especially during slow-speed turns or steep climbs under conditions when the actual horizon is not available for reference, it is a totally reliable means of keeping the pilot from inadvertantly getting himself into a stall or worse, stall-spin. The advantage of such an instrument *cannot* be exaggerated and, being reasonably priced, any aircraft owner with *any* sense should consider having one installed.

Lights

Aircraft are equipped with navigation lights, those red, green and white lights on the wingtips and tail, an anti-collision light which may be a red rotating beacon or a strobe, a landing light, and cabin lights. All these lights are controlled by switches, usually rocker switches, most often located at the bottom portion of the instrument panel. The navigation lights and the anti-collision light must be turned on when flying at night. The landing light is turned on during landing and while taxiing on the ground. For the benefit of those who have never made a night landing it should be pointed out that, except during the last few seconds prior to touchdown, the landing light serves no practical purpose other than to make it easier for the tower and other aircraft to see the airplane. It does *not* effectively light up the runway, and therefore cannot be used to take the place of runway lights during a landing in the dark. But a landing light (or, if available, taxi lights) should always be used when taxiing on a dark airport. There have been cases when the wind has blown some debris onto the taxiway, or line personnel have been known to leave such things as aluminum ladders standing around somewhere, and it could be an expensive pleasure to have the prop collide with some such obstruction because we simply were unable to see it until it was too late.

There are two types of cabin lights. There is a white dome light which simply illuminates the entire cabin, and which is rarely used in flight. And then there are the lights which illuminate the instrument panel and a certain area of the cockpit where charts may have to be studied. In most cases, these lights can be adjusted by the pilot to burn either red (or blue) or white, and they are on a dimmer so that their intensity can be adjusted. In addition, some instruments are equipped with either internal or so-called *post lights* which are also on a dimmer and can be left on when all other illumination in the cabin is extinguished. The amount of interior light to be used during night flight is a matter of individual preference. Some pilots like to use an absolute minimum, while others are perfectly happy to fly around lit up like a Christmas tree.

Temperature Controls

Aircraft are equipped with adjustable vents, heaters and defrosters. The vents bring fresh air into the cabin and, when intelligently designed (which is not always the case), they can be adjusted to direct that fresh air to specific portions of the cabin. The heater is an important factor, considering that much of the time the air at altitude is pretty cold. But heaters occasionally spring leaks which could result in small amounts of carbon monoxide being funneled into the cabin along with the heated air. Since carbon monoxide has neither smell, taste or color it is undetectable. Therefore, whenever the heater is being used, some of the fresh air vents should be kept at least partially open.

The defroster works just like the one in an automobile. It directs heated air against the inside of the windshield. There are two situations during which a defroster can become important. One, and the most frequent, occurs when we climb into an airplane when it's been raining. Our breath and body temperatures tend to cause the windshield to become fogged over, reducing visibility to zero. The other, less frequent but more serious, happens during the winter when sudden snow or ice accumulations can deprive us of forward visibility in flight. As long as the defroster is started early and the snow or ice condition is not too serious, it will do its job. But if it is turned on too late, or if the conditions are really serious, it cannot be relied upon to be very effective.

Flight Controls

The aircraft controls don't really fall into this category, but being part of all that a pilot has to deal with in the cockpit, we might briefly mention them here.

First there is the control wheel, also referred to as *yoke* or *stick*. It operates the ailerons and the elevator and thus is used to determine the direction of flight by controlling bank, and the airspeed by controlling the angle of attack. While it is usually and incorrectly also thought of as controlling altitude, it does so only in conjunction with the appropriate adjustments in engine power. By itself it is incapable of producing a sustained climb or controlled descent.

Then there are the rudder pedals, which also include the wheel brakes. In most aircraft the rudder pedals are attached not only to the rudder but also to the steerable nosewheel. Thus, on the ground, the airplane is steered during taxi by use of the rudder pedals. Pressure on either side will initiate a turn in that direction. Pressure with the toes on the upper portions of the rudder pedals will activate the wheel brakes on the main wheels. Since the toebrake on the right pedal activates the brake in the right main wheel, and the left one the other, it is possible to use differential braking in order to achieve a really tight turn. (Aircraft with free turning nosewheels must always be steered by differential braking until sufficient speed has been reached to make the rudder effective.)

The trim wheel or crank is used to adjust the trim of the elevator (and in some aircraft the rudder), as has been discussed in some detail earlier.

And then there is some type of lever which, either mechanically, hydraulically or electrically, controls the position of the flaps. Some flaps can be deployed and stopped at any point during the flap travel, while others can only be stopped at certain points, such as 10 degrees, 20 degrees and 40 degrees, or some such. Personally I have always liked the manual flaps which used to be operated by a huge handle in the older Cessnas best. Somehow they seemed to provide the pilot with better control, but like so many of the older, simpler (and foolproof) systems, manual flap controls seem to be a thing of the past.

And last, if the aircraft is equipped with a retractable gear, there is the gear lever or handle which is used to retract and extend the landing gear. There is a saying among pilots of retractable-gear aircraft that if you have not made a belly landing, some day you will. To reduce the chance of that happening, aircraft are equipped with gear-warning devices, loud blaring horns and lights on the instrument panel, which are supposed to alert the pilot that it is time to put the gear down. (In addition, tower controllers have, in recent years, added the phrase "Check gear down" as a matter of routine to all landing clearances.)

So much for the instruments and controls found in an aircraft cockpit which are not, in one way or another, related to the operation of the engine. So now let's take a look at that group.

Master Switch

The master switch simply activates the electrical system by connecting the battery with the lines leading to all electrical instruments, systems and devices. It may be a separate switch, or it may be combined with the left-right-both selector.

Left-Right-Both Selector and Starter

This is usually a rotary switch with five positions: *off, on, left, right* and *both*. In the *off* position nothing happens. When turned to *on* it activates the starter motor which turns the engine and the prop and gets it started. Then later, prior to takeoff, we place it into the *left* position in which only the left magneto operates its one set of spark plugs. There will be a slight drop in rpm, usually not more than 50 rpm or so. If the drop is beyond the acceptable limit for that particular engine, then there is something wrong and flight should not be undertaken. If it checks out all right, we now switch to *right* to test the other magneto and its set of spark plugs. If that, too, checks out okay, we switch back to *both* which means that both magnetoes and both sets of spark plugs will be used, and the switch will then remain there throughout the flight.

Electric Fuel Pump

Most aircraft are equipped with an auxiliary, usually electric, fuel pump. On some, this pump must be used every time the engine is being started. On others it should not be used except during certain emergencies. Information with reference to the use of the auxiliary fuel pump is found in the Pilot's Operating Handbook, and the pilot should familiarize himself with that information before attempting engine start.

Throttle

As in an automobile, the throttle is used to control the amount of fuel-air mixture which is fed to the engine and, in turn, the amount of power produced by the engine. In older light aircraft (except twins) all throttles used to be of the plunger type, some simply to be shoved in and out, while others could be turned for better control of small adjustments. In recent years, aircraft manufacturers appear to have decided that the type of lever throttle used in twins is sexier, and today many single-engine aircraft are equipped with that type of engine-control pedestal which uses levers instead of plungers. The effect is exactly the same, except that I have always found that minute adjustments can be made more easily with the plunger-type controls.

In all throttle installations, moving the throttle forward increases power and pulling it back reduces power. But there is a difference in throttle action between aircraft with fixed-pitch props and those with constant-speed propellers. In aircraft with fixed-pitch propellers, the throttle controls the rpm. (Actually it controls manifold pressure which, in turn, governs the speed with which the engine turns the crankshaft and, in turn, the propeller.) Here, full throttle will automatically result in maximum available rpm. In aircraft with constant-speed propellers, the throttle adjusts only the manifold pressure and has no effect on the propeller rpm (except when manifold pressure is drastically reduced). The rotation speed of the propeller itself is adjusted by the pilot by using a separate control.

A dial on the instrument panel, usually marked MP, shows the pilot the amount of manifold pressure which is available at any given moment and throttle setting. It is calibrated in *in hg*, and, especially in turbocharged aircraft, there will be a red line, indicating that the manifold pressure must never exceed the number which coincides with that red line. As a general rule, we use all the manifold pressure we can get during takeoff, and then reduce it somewhat during climb to save fuel and to keep the engine from overheating.

Propeller Control or RPM

This control, which usually looks pretty much like the throttle, is found only in aircraft with constant-speed propellers, and is nearly always located right next to the throttle. (For some dumb reason it is not always on the same side of the throttle, thus creating confusion when pilots have to switch from one type of aircraft to another.) Pushing the lever forward will increase propeller rpm, and pulling it back will reduce rpm. The actual rpm are displayed in a cockpit instrument marked *rpm* and calibrated on 100s of revolutions per minute (Fig. 5-15). Thus, a 25 indication means that the propeller is turning at 2,500 rpm.

It might be appropriate here to dispel a myth about the numerical relation between manifold pressure in inches of mercury and the rpm of the propeller, which has resulted in the insistence by some pilots on always operating with a so-called squared power setting, such as 25 inches of manifold pressure and 2,500 rpm. That is utter nonsense. There is *no* relation between the two, except

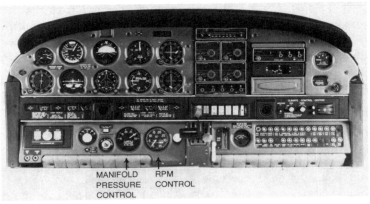

MANIFOLD   RPM
PRESSURE   CONTROL
CONTROL

Fig. 5-15. The manifold-pressure and rpm readout instruments in this aircraft are at the bottom left center of the instrument panel (courtesy of Piper).

that using excessively low rpm with high manifold pressures can damage the engine. For this reason, it is a hard and fast rule that when increasing power you first increase rpm and then increase manifold pressure, and when reducing power you first reduce manifold pressure and then rpm. It is also the reason why the prop should always be set to high rpm before landing. Then, in the event of having to go around, a sudden increase in the manifold pressure will not overstrain the engine.

Mixture Control and EGT

Mixture in this context refers to the ratio of air to fuel which is fed to the engine. While in all modern automobiles the mixture is fixed and cannot be adjusted by the driver, in aircraft it is up to the pilot to adjust the mixture, an adjustment which is vitally important because of the different altitudes at which the aircraft operates.

Mixture is controlled by another lever or plunger which looks much like the throttle, but usually has a red head or knob in order to be easily distinguished at a glance. But unlike manifold pressure and rpm, there is no readout in the cockpit which tells the pilot what mixture he is using, except in aircraft equipped with so-called exhaust-gas-temperature gauges (EGT).

Mixture is such an important and frequently misunderstood subject that we'll be devoting an entire chapter to the subject. The correct use of mixture control will extend the life of the engine. It can cut fuel consumption by huge amounts and, in turn, extend the range of the airplane by a meaningful percentage.

The fuel-air ratio—in other words, mixture—controls the temperature of the exhaust gases. A rich mixture will result in relatively cooler exhaust gases than will be produced by a leaner mixture. But once a certain amount of leaning, known as *peak*, is exceeded, the exhaust-gas temperatures begin to drop again. That is the principle on which EGTs work. A probe is placed in the exhaust manifold. It senses the temperature of the gases and displays this temperature on a readout in the cockpit. With such an instrument, the pilot can lean the mixture until the needle hits its highest reading and then starts to move down again (Fig. 5-16). The highest reading represents peak. Some engines may be operated at peak, some on the lean side of peak, while others require that they be operated on the rich side of peak. But more about this in the chapter on mixture.

Some EGTs, so-called engine analyzers, have probes in the exhaust manifold for each cylinder. By being able to compare the readings related to each cylinder, it is possible to detect abnormalities and thus to anticipate possible engine troubles. EGTs and engine analyzers are not particularly expensive (relative to what everything in aviation costs) and should be considered a necessary piece of equipment in every airplane (Fig. 5-17).

Carburetor Heat

This is another one of those controls which is peculiar to aircraft, to be specific, only the aircraft with carbureted (not fuel injected) engines, but which is frequently misused by pilots. Here is what it's all about: The action of the carburetor which mixes fuel and air for use by the engine, for some reason which is not important to be explained here, results in a drastic temperature drop. When there is no appreciable amount of moisture in the air

Fig. 5-16. The cockpit readout of an exhaust-gas-temperature gauge (courtesy of Alcor).

Fig. 5-17. The cockpit instrumentation and associated hardware of an engine analyzer for a four-cylinder engine (courtesy of Alcor).

(and there is not supposed to be any detectable amount of water in the fuel), this phenomenon is of no consequence. But when the air is heavily laden with moisture, the moisture tends to freeze, creating some solid ice in the carburetor throat. If this ice buildup is permitted to continue for too long, it will gradually shut off the fuel supply to the engine, eventually starving it and causing it to quit.

The carburetor-heat control, when activated, sends hot air to the carburetor which melts the ice and permits continued fuel supply for the engine. Carburetor icing conditions can develop on the hottest of summer days. They have little to do with the outside air temperature. Whenever the outside air temperature and the dew point are within a few degrees of one another, the situation is ideal for the formation of carburetor ice. (The dew point is the temperature to which the prevailing temperature has to drop in order for the air to become saturated with moisture or, in other words, to produce 100 percent humidity. When the air is dry, the spread between current temperature and dew point is wide. When the humidity is high, the spread between the two is narrow.)

The first sign of the beginning of accumulation of carburetor ice is a gradual drop in engine performance. When this happens without any other logical explanation, it is time to activate the carburetor-heat control. To do this we *pull out* the carburetor-heat plunger *all the way*. (The carburetor-heat control works opposite to the way the throttle and other controls work. Pulling it out increases its action. Pushing it in reduces its effect and eventually stops it altogether.) Carburetor heat should *never* be partially applied. It's a question of all or nothing. When it's needed, use *full* carburetor heat. When it's *not* needed, turn it *fully* off. But once the need for carburetor heat has been detected, put it on all the way and *keep* it on all the way until it has become quite obvious that the condition which was conducive to the accumulation of carburetor ice is no longer.

The application of carburetor heat will reduce engine power to a minor degree and it will cause an automatic enrichment of the mixture being used. Thus, when carburetor heat is used, the mixture should be adjusted to the same degree of leanness which was used before. And once the use of carburetor heat has stopped, the mixture must again be adjusted in order to avoid operating too lean.

Alternate Air

Without air, an engine can't function. But there is always the chance, though slim, that under certain weather conditions the air intake could be covered with snow or ice. To avoid the inevitable result of complete engine failure, many aircraft are equipped with an alternate-air control. When this control is moved to the alternate-air position, air from within the cabin or other protected area is fed to the engine. The resulting engine power will be less because air fed through the regular intake is rammed into the intake at considerable force because of the speed of the airplane, and this ram-air effect is missing when the alternate-air source is used. Still, at least it keeps the engine from quitting.

Fuel-Flow and Fuel-Pressure

Most, though not all, aircraft are equipped with a readout instrument which shows either the fuel pressure in terms of pounds per square inch (psi) or in terms of fuel flow calibrated in either pounds or gallons per hour. The gauges showing fuel pressure tend to be more reliable, except that we don't really care about fuel pressure except when it suddenly and inexplicably drops to below the acceptable minimum. Fuel flow, on the other hand, the amount of fuel being burned per hour, is something we *do* like to know, except that gauges showing these parameters can be relied upon to be unreliable. They are *approximations*, at best.

The only reliable instrument showing the amount of fuel being used which has yet been devised is a piece of computerized optional equipment that, because of its fairly hefty price tag, is not found in too many light aircraft. The most popular such instrument is called Fueltron. It works like this: Before takeoff you dial in the amount of fuel on board. Then, from the moment the engine is started, the system will keep track of every drop of fuel being used, and at any time during the flight it can present the following data to the pilot: fuel burned since takeoff; amount of fuel remaining; amount of fuel being burned in terms of gallons or pounds per hour at the current power setting; time remaining on the available fuel assuming no change in power setting. It's a marvelous piece of equipment and completely reliable, and anyone who can afford its $1,500-plus price tag (for single-engine aircraft) could probably recoup its cost in fuel savings over a period of somewhere between 500 and 1,000 hours of flying.

Fuel Gauges

All aircraft are equipped with gauges which show the amount of fuel in each tank. The trouble with these gauges is that they appear to be inherently unreliable. Some will show a small amount of fuel left when, in fact, the tank is empty, while others will show empty long before they actually are. For this reason, fuel gauges should *not* be relied upon except as an approximation of the amount of fuel on board.

A word of warning. Fuel gauges are electric gauges which respond to the information provided by a so-called sending unit in the fuel tanks. If, for some reason or another, the electrical system should malfunction, they will simply read empty, no matter how much fuel is in the tanks.

Cylinder-Head-Temperature

The cylinder head, the upper portion of the cylinder in which all those explosions take place, is the hottest part of the engine, and being able to monitor its temperature provides the pilot with an opportunity to avoid overheating the engine. For this reason many (not all) aircraft are equipped with cylinder-head-temperature gauges which are redlined at the never-exceed temperature. Though by no means a necessity, it's a good thing to have, and careful attention to the cylinder-head temperatures will most probably result in a meaningful extension in engine life.

Oil-Pressure and Oil-Temperature

These gauges do just what the name implies: They show the amount of oil pressure and the temperature of the oil. An engine cannot function without a continuous supply of oil, and this supply must be under a given amount of pressure in order to accomplish its lubricating function. If either gauge shows an abnormal reading, it's probably time to land and have a mechanic look at the engine. But the readings of the two gauges are related to one another. If the oil-pressure reading should suddenly drop, check the oil temperature. If it remains perfectly normal, then it is safe to assume that the trouble is in the gauge or its monitoring system, because if pressure had actually dropped, the temperature would have risen rapidly. On the other hand, if the oil temperature is going past the redline, regardless of the pressure reading, an immediate landing is indicated.

As a general rule, aircraft should not start their takeoff run until the oil temperature has risen into the green. Still, on very cold winter days it is sometimes impossible to get the temperature into the green prior to takeoff. In normally aspirated engines this is usually not too serious and it is standard procedure to take off and let it creep into the green during the climbout. In turbocharged aircraft this is not recommended. With the turbine in the turbocharger rotating at more than 100,000 rpm, lubrication is critical, and oil which is still too cool may simply be too thick to do the job.

Ammeter

This instrument simply shows that the generator or alternator is functioning properly. It should read *charge* after heavy use of electricity, such as after engine start, but should return to normal when there is no excessive current drain. If it shows *discharge* for an abnormally long time, it can be assumed that there is a short or, at least, some frayed insulation somewhere. If it remains on *full charge* for an abnormally long period of time, it is probably an indication of a malfunctioning voltage regulator. When this happens in flight, it is safer to simply turn off the master switch and continue on without the electrical system, because continued operation under those conditions can result in the battery either simply melting away or, worse yet, exploding.

Suction Gauge

All gyroscopic instruments, the directional gyro and the artificial horizon, depend on suction for their operation. In times past, this suction was frequenty supplied by a venturi tube attached to the outside of the fuselage. But suction provided by such a device is unreliable because of the variations in airspeed. Therefore, modern aircraft are equipped with engine-driven suction pumps. Since excessive fluctuations in the available suction will cause the gyro instruments to become unreliable, a suction gauge is part of the instrumentation, providing the pilot with an opportunity to make sure that the suction remains constant.

Hobbs Meter

Aircraft used primarily for training or rental are equipped with a so-called Hobbs meter. This instrument counts the hours of operation, starting with the moment the master switch is turned on, and stopping only when it is turned off. Its reading does not coincide with the hour-reading in the rpm display, because that reading is based on actual revolutions per minute. Therefore, the Hobbs-meter reading is always higher than that which displays engine hours.

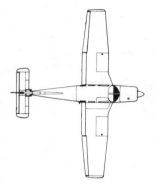

Chapter 6
Aircraft
Performance
and Limitations

Every aircraft is designed to operate within certain performance parameters and limitations. While the actual figures for each of these parameters and limitations vary from aircraft to aircraft, a quick look at what is meant by each of these terms is helpful in understanding the information provided in the different Pilot's Operating Handbooks.

Takeoff

The takeoff parameters consist of *takeoff ground run; distance to clear a 50-foot obstable;* and *liftoff speed.* (In jet aircraft the latter is divided into rotation speed—V_1—and liftoff speed—V_2.) The takeoff ground run is the distance it will take the aircraft, at full power and full gross, to accelerate to liftoff speed on a smooth hard-surface runway under no-wind conditions, at sea level on a standard-temperature day. Since very few takeoffs are ever made at sea level at standard temperature, the published minimum ground-run distance should always be increased by at least 50 percent in order to be on the safe side. Another reason for this precaution is the not illogical assumption that your own pilot technique is not up to that of the company test pilot. The newer versions of the Pilot's Operating Handbooks often also include takeoff ground-run distances under such conditions as grass strips, wet runways or varying wind conditions. This helps a lot, but still, they are minimums and should be accepted as such.

The distance to clear a 50-foot obstacle is just that. The absolute minimum distance in order to climb to 50 feet agl, again using the assumed conditions stated above. Here, too, a *generous* margin of safety could certainly be considered the better part of valor.

Liftoff speed is the *indicated* airspeed at which the airplane is capable to initiate and then sustain flight. Depending on the airport elevation and the prevailing temperature, the actual ground speed or true airspeed may vary greatly, but the indicated airspeed will always be the same. Under conditions of no wind or a straight-on-the-nose headwind, liftoff at this minimum speed is usually quite safe and saves wear and tear on the tires. But under crosswind conditions, it is more often advisable to hold the aircraft on the ground until a somewhat higher indicated airspeed has been reached.

The liftoff speed varies depending on the use of flaps. It will be considerably higher if little or no flaps are used, and is usually lowest with the recommended (usually half) flap setting. Using more flaps tends to increase the amount of drag and increase the amount of runway distance needed in order to achieve the necessary liftoff speed.

For special techniques used for takeoff from (or landing on) short fields, check the special chapter on that subject.

Climbout

An aircraft can be climbed at a variety of speeds. The steepest, the one that reaches the most altitude while covering the least distance horizontally is known as the *best angle of climb*. It is uncomfortable for pilot and passengers alike because the nose of the airplane points way up into the sky resulting, from the pilot's point of view, in very poor visibility. But there are times when it may be important, such as when much altitude must be obtained quickly in order to clear rising terrain. The best-angle-of-climb speed varies from airplane to airplane and is listed in the Pilot's Operating Handbook. At this speed the aircraft operates uncomfortably close to the stall speed, flying at a very high angle of attack.

There are times when we may be tempted to use this steep angle of climb in order to quickly get up to the level of smooth air when it's annoyingly bumpy down low. This is *not* to be recommended because bumpiness, being the result of up- and down-drafts, could easily bring us momentarily to the edge of the stall, something which we would certainly want to avoid except when operating at a safe altitude.

The other published climb speed is known as the *best rate of climb*. This speed is the result of a combination of speed and power which accomplishes the greatest altitude gain in the least amount of time. It results in a somewhat shallower angle of attack and covers more distance horizontally. At this speed, the nose of the aircraft is somewhat lower than at the best angle, but it still tends to block out much of the pilot's field of vision.

During periods of high outside-air temperatures, both of these speeds, being relatively slow while the engine operates at full or nearly full power, may result in engine overheating because the insufficient airflow produces less cooling than is needed.

The most popular type of climb is the so-called *cruise climb*, an expression which stands for any speed at which the airplane climbs at a higher forward speed and a lesser rate of climb (and at a shallower angle of attack) than in either of the two previously described. The advantages of cruise climb are better visibility for the pilot, a more comfortable aircraft attitude for the passengers, and improved engine cooling. Cruise climbing means that it takes a little longer to get to the desired altitude, but a lot more horizontal distance is covered while getting up there.

Much has been written about which type of climb results in the greatest fuel efficiency over the entire distance of the flight. Some people will argue that it is cheaper to get up to altitude in a hurry, because that results in a longer time in cruise at a reduced power setting. Others will be convinced that cruise climbing, which usually is accomplished at a less-than-full-power setting, is the better deal in the long run. The fact is that it makes *no appreciable difference*. Regardless of the speed and angle of climb, the total time in flight and the total amount of fuel burned will be just about the same.

Cruise

This is where we spend most of our time and this is where intelligent fuel and speed management will bring the most meaningful results. Cruise speeds are normally described in terms of percentages of power. Thus we will say that we cruise at 75 percent or 65 percent or whatever. In the "good old days" when aviation gasoline was relatively cheap, we didn't tend to worry too much about the amount of fuel burned, and since the primary reason for flying is to get somewhere in a hurry, we'd automatically use the power setting which would result in the highest airspeed. This usually meant operating at 75 percent of power at low altitudes, and at full throttle at any altitude at which full throttle would produce 75 percent or less.

Today, with aviation fuel costing in excess of $2 per gallon, all this has changed. Suddenly, economy has become as important or even more important than speed. In the process we have begun to realize that a slight reduction in the percentage of power being used can save huge amounts of fuel while producing only minor penalties in terms of speed. For example, an aircraft cruising at 10,000 feet at a power setting which produces 76 percent of power will move along at a TAS of 159 knots (183 mph), burning 14.5 gallons per hour. That same aircraft flying at that same altitude at a power setting resulting in 53 percent of power will achieve a TAS of 135 knots (155 mph) while burning only 10.3 gallons per hour. Assuming that the distance to be covered in cruise is 400 statute miles, it will take two hours and 12 minutes at the higher speed, costing 31.7 gallons of fuel. At the lower speed it will take two hours and 36 minutes at a cost of 26.6 gallons. At $2 per gallon, the 24 minute difference would cost us $10.20. There may be times when this is worth it, but most of the time it's not. And if you spread this type of arithmetic over a year's flying, the savings which result can easily pay for most of the maintenance costs. (The figures used here apply to a 1980 Cessna Turbo Skylane RG, Fig. 6-1).

But the percentage of power to be used is not the only consideration affecting cruise. The others are altitude and the prevailing winds at the various altitudes. As we will see in a later chapter which concerns itself with fuel economy in some detail, it is more economical to fly high than low, and, depending on the velocity of the head- or tailwind components, it may be advantageous to use a higher true airspeed when flying into the wind, and a lower one when the wind is on our tail.

Fig. 6-1. Cessna Turbo Skylane RG (courtesy of Cessna).

While on the subject of cruise, every airplane has something called *service ceiling* and something else referred to as the *absolute ceiling*. These two are not the same, and they vary somewhat with airplane weight. The service ceiling, a figure which is published by the manufacturer for each airplane, is the altitude at which the aircraft, at full gross and using all available power, is still capable of climbing 100 fpm. The absolute ceiling is the altitude at which, at a given weight, the airplane simply refuses to climb any higher. At that altitude it will be mushing along at an IAS which is not much above or even equal to the best-rate-of-climb speed, though the TAS may be quite considerable. (At 20,000 feet with an outside air temperature of minus 5 degrees C. an IAS of 100 mph equals a TAS of 142 mph.) But for most aircraft the absolute ceiling is so high as to be of no practical consequence.

The service ceiling, on the other hand, is something to keep in mind when we have to fly over some of the higher mountain ranges or are trying to get over the tops of some cloud buildings. The ability to climb at 100 fpm means that accomplishing a 1,000-foot altitude increase will take 10 minutes, and after that, the higher we go, the less the rate of climb will be. It would therefore seem not particularly sensible to approach a 14,000-foot mountain range in an airplane with a 13,000-foot service ceiling. I don't say that it *won't* get you across all right, especially since the airplane is not likely to be at gross by then, but it'll be marginal *at best*.

Descent

Descents, like climbs, can be either steep or shallow. But unlike climbs, during which the gravity pull of the earth produces a negative effect, this same pull gives us a free assist as we're trying to lose altitude. And in aviation it is always a good idea to take advantage of all the free assistance we can get. Therefore, whenever weather conditions (or air traffic control) permit, a long and gradual descent is more fuel efficient, especially if we resist the temptation to increase speed, than is a steep one. The important thing here is to pick the rate of descent we like, say, 300 or 400 fpm, and then to figure out when the descent has to be started in order to get us to pattern altitude at the destination airport by maintaining the same speed which we have previously used in cruise. (An airplane traveling at a ground speed of 150 mph and descending at 300 fpm will cover 83.3 statute miles while losing 10,000 feet of altitude.) When making these calculations it is important to think in terms of ground speed, not true airspeed, because what we are interested in is the point on the ground over which we have to start our descent.

The most wasteful situation in terms of descent is one in which we find ourselves over the destination airport with thousands of feet of extra altitude. The time and fuel spent to come down from there are a complete waste. In addition, a rapid descent with the engine at idle tends to produce a very rapid cooling effect on the engine which is bad business and which, in extreme cases can cause the engine to refuse to respond when power is again needed during the approach. Under the worst of circumstances, it can actually result in engine damage.

In view of all of the above, a descent requires more careful preplanning than does the average climb.

Landing Distances

Like takeoff distances, landing distances are divided into the distance required to land over a 50-foot obstacle and the actual ground run after touchdown. Again, except in pilot's handbooks in which these distances are given for a variety of weather and airport-elevation conditions, the published figures assume sea level and standard temperatures plus a dry paved runway. They also assume a high degree of pilot technique and good brakes. Here, too, it is therefore advisable to add a good 50 percent "fudge factor" to those published distances and consider that the *actual* minimum.

It is a generally accepted fact that the landing is the most dangerous and difficult part of flying. It is dangerous because we are close to the ground, and the contact between the airplane and the ground must be just right in order to avoid some type of mishap. It is difficult because no two landings are ever exactly alike. In addition, landings are always made at the conclusion of a flight, and if that flight has been of any length, the pilot is likely to be somewhat more tired and therefore less proficient than he was at the time of takeoff. In addition, many landings must be made at unfamiliar airports, and that tends to add to the problem. Because landings can be so varied (and they are, in fact, unavoidable), we will devote an entire portion of another chapter to that subject. Suffice it to say here that since landings and the maneuvers immediately prior to landing must be made at low speeds, any pilot should practice slow flight (at a safe altitude) until he feels comfortable and proficient at it.

Other Limitations

Every aircraft has what is referred to as an *accelerate-stop distance*. That is the distance required to accelerate an aircraft from a standing start to liftoff speed and, assuming failure of the engine (or, in twins, the critical engine) at the instant that speed is reached, to bring the aircraft to a stop, using heavy braking. Quite obviously, the accelerate-stop distance for any aircraft will vary considerably under different wind and runway-surface conditions.

Older aircraft owners manuals never listed the accelerate-stop distances. The new type of Pilot's Operating Handbook does list it for most aircraft. While we may fly thousands of hours without ever having to abort a takeoff, the possibility is an ever-present one. For this reason pilots should think twice about accepting an intersection takeoff. The full length of the runway may be a great deal more than we feel we need, but if, for one reason or another, we should find that the takeoff has to be aborted, it would be depressing indeed to find ourselves running off the end of the runway and into a fence. As the saying goes, there is nothing more useless than the concrete *behind*.

Instead of an accelerate-stop distance, jet aircraft have something called *balanced field length*. The technical description of balanced field length is the distance within which a jet aircraft, under given weather and runway conditions, can accelerate to V_1 (rotation speed) and then either stop or accelerate to a safe climb speed (V_2) and clear a 35-foot obstacle on one engine.

Chapter 7
Operating the
Flight Controls

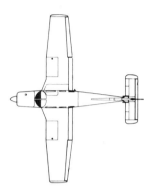

In order to attract new flight students, aircraft manufacturers and flight schools will occasionally say in their advertising that anyone who can drive an automobile can also fly an airplane. Aside from the fact that this is not really true, it tends to produce a mental picture of an airplane as a flying automobile, and that is not only completely false, but can actually be *dangerous*. Both automobiles and airplanes are transportation machines, but that is where the similarity stops. Unlike the automobile, the airplane operates in a three-dimensional environment and the manner of controlling its flight path is *entirely* different.

Most students, at one time or another during some of the early hours of flight—more often than not during the first cross-country solo—suddenly realize with a degree of shock that in an airplane you can't "pull over" in order to sort out what needs to be done next. In addition to the difference in the operating environment, the fact that an airplane *can't* stop and wait makes it an entirely different sort of animal. In an automobile going slow is equivalent to being safe. In an airplane "low and slow" spells danger, while "high and fast" is what safe flying is all about.

In other chapters we have talked at length about the angle of attack, the attitude of the airplane relative to the air within which it is flying, which is what makes flight possible in the first place. Now let's take a closer look at the controls, and how they must be used by the pilot, in order to make the airplane do what it is supposed to do. The best way to do this would seem to be to take the airplane through every possible flight condition and to mentally watch the pilot as he handles the controls.

Preflight, Startup and Taxi

It is the pilot's responsibility to preflight the airplane but, at first at least, being used to climbing into our automobiles and driving off without giving it another thought, we tend to feel kind of silly as we crawl around, pushing and pulling various parts of the airplane. It requires the psychological realization, rarely enunciated by flight instructors, that if something turns out to be amiss after we're airborne, it's a bit late to do anything about it. In other words, we'd better make sure that the airplane is in tip top shape before we entrust our lives to it.

This then is the purpose of the preflight. It involves an examination of all moving surfaces of the airplane to reassure ourselves that there are no loose screws or missing rivets or some such, and that there is nothing untoward to inhibit the full travel of each moving part (Fig. 7-1). We check the oil to make sure that there is an ample supply. And we drain the sumps because moisture collects at the bottom of the tanks and it should be gotten rid of. A tiny amount of moisture is nearly always present in the fuel and causes no real problem. But a cup full of water drawn into the engine would more likely than not result in complete and sudden engine failure.

We may, at times, see a pilot do what appears to be a very cursory preflight inspection. That does not necessarily mean that he is sloppy or careless. Years of experience with a variety of airplanes teach us what to look for at a glance. There is an old joke about a preflight being a quick look to make sure that "nothing is hanging or dripping." but that's all it is, a joke. Any pilot who has respect for his life and the life of his passengers will do a *thorough* preflight, even though it may not always be obvious to the casual bystander.

Okay, nothing is hanging or dripping, so we climb aboard and strap ourselves in. The first thing to do now is to adjust the seat so that the rudder pedals can be easily reached. The next move is something that should *always* be done and rarely is: Make sure that the seat is firmly in place and can slide neither forward nor backward. In many airplanes the seats can be moved backward so far that reaching the controls becomes impossible. If that happens inadvertently during acceleration or climbout, the result could be

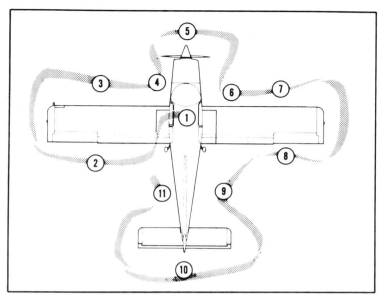

Fig. 7-1. Diagram of a routine preflight walkaround (courtesy of Beech).

catastrophic. The pilot would practically automatically grab for the wheel in order to pull himself forward. This would cause the wheel to be hauled all the way back, pulling the nose of the airplane steeply up into the sky, making matters worse and resulting surely in a stall. That kind of an uncontrolled departure stall can't help but result in a pile of bent aluminum, not to mention the condition of the pilot and anyone else on board.

We're strapped in and the seat is firmly in place. We now make sure that the parking brake is set and, as an added precaution, we push on both toe brakes. After making sure that no one is standing too near, we yell "Clear" through the open storm window (or open door) and start the engine. The starting procedure varies among airplanes and there is no point in going into details here. Once the engine has settled down to a comfortable idle we turn on the radio and contact ground control (assuming this is a controlled airport):

"Big City Ground, this is Cessna One Two Three Four Zulu at Alpha Aviation. Taxi to takeoff."

Ground control will come back with the necessary taxi instructions, the active runway, the altimeter setting and any other information which we are supposed to be aware of. We'll now set the altimeter to the airport elevation and the numbers in the adjustment window should coincide with the altimeter setting reported by Ground Control. If it does not, there is something wrong with the altimeter and while, except in extreme cases, this would not be sufficient reason to abandon the flight, we might make a mental note to have it checked at the next opportunity.

We now release the parking brake and toe brakes and start to taxi. On the ground, an airplane is steered by the rudder pedals. New students tend to automatically turn the wheel when trying to make a right or left turn which is an exercise in futility. It takes a bit of getting used to, to steer with the feet, but it doesn't take long. There is no need for the pilot to keep his hands on the wheel at all except under certain circumstances. When there is a strong crosswind it is a good idea to turn the wheel toward the wind. This will cause the upwind aileron to be raised, reducing the amount of lift produced by that wing and thus minimizing the chance of the wind lifting that wing and causing an upset. Also, when taxiing downwind, if the wind is strong, it is a good idea to hang on to the wheel to prevent having the wind slam the elevator up and down. Another situation is one where we have to taxi across extremely rough ground. In that case, using a certain amount of back pressure (back pressure means pulling the wheel toward us) will reduce the weight on the nosewheel and thus minimize the chance of damage to it.

Once having arrived at the runup area adjacent to the takeoff end of the active runway, we now face the airplane into the wind and again set the parking brake. We're now ready for a series of last-minute checks for which flight instructors have invented the term *cigar* as a means of remembering what should be checked: C=controls; I=instruments; G=gasoline; A=attitude; R=runup.

Controls: We operate the control wheel through its full travel in all directions and, and this is important, by looking out make sure that the control surfaces respond the way they are supposed to. Turning the wheel to the right should raise the right aileron and lower the left one. Pushing it forward should lower the elevator. Pulling it back should raise the elevator. This visual check is vital because there have been instances where the control cables were inadvertantly reversed by some not-quite-with-it mechanic, or where a control cable simply broke or was disconnected. So, just the fact that we can move the wheel is no proof that it is producing the desired effect. A visual inspection is the only foolproof method. We now run the flaps all the way down and then raise them again to the setting which is recommended for takeoff in this particular airplane. (Except when there is going to be a strong crosswind during the takeoff run. In that case we may want to use no flaps at all. Flaps tend to increase the adverse effect of the crosswind on the controllability of the airplane, and a no-flap takeoff, assuming sufficient runway length, is likely to be preferable). In airplanes in which the rudder is not connected to the steerable nosewheel, we can now also check the rudder by pushing on one pedal and then the other and by leaning far out of the door, making sure that right rudder turns the rudder to the right (looking theoretically at the airplane from the back) and that left rudder turns it to the left. In aircraft in which the rudder and nosewheel are interconnected this is a little more difficult because turning the nosewheel with the airplane not in motion puts considerable strain on the connecting cables or rods.

Instruments: Unless the altimeter was set earlier (as it should have been), it must be set now. The directional compass should now be aligned with the magnetic compass. The clock should be set to the correct local or Zulu time, and the time be noted as being the approximate takeoff time. The transponder should be set on standby and, if this is a VFR flight, should be set to 1200. The nav radio should now be tuned to the first VOR to be used for navigation, and the com transceiver must be tuned to the tower frequency. And, depending on the amount of avionics in the airplane, other radios should be tuned to the frequencies to be used early in flight.

Gasoline: This is simply a reminder to check the fuel selector and make sure that it is on the fullest tank or, in aircraft so equipped, on "both." The fuel quantity should have been checked earlier during the preflight by visually checking each tank.

Attitude: This refers to the setting of the trim tab. The trim-tab-setting indicator should now be moved to the position marked for takeoff. This is important because if the trim tab is accidentally left in an extreme nose-high or nose-low position it would tend to create control problems immediately after liftoff. Since the trim is usually in a nose-high position for landing, it is likely to still be in this position when we're getting ready to start the next flight, and remembering to adjust it before takeoff could prove critical.

Runup: The runup procedure differs depending on whether the aircraft is equipped with a fixed-pitch or constant-speed propeller. In aircraft with fixed-pitch propellers, we simply advance the throttle until the recommended rpm is reached. We then move the ignition key first from "both" to "right" (or "left") and observe the drop in the rpm. If that drop is within the recommended limits we then move it back to "both" and then to "left" (or "right") and again check the rpm drop. Assuming that too is within the allowable limits, we return the key to "both" and pull the throttle back to idle. If the aircraft is equipped with a constant-speed prop, we put the prop into the high-rpm position and then follow the same procedure described above. Then, after the key has been returned to "both," we pull the rpm control back to the low-rpm setting for a brief moment and then advance it again all the way. This should be done two or three times (more if the outside temperature is very cold) in order to loosen the oil in the control mechanism. This action is generally referred to as "exercising the prop."

If the takeoff is to take place from an airport with a high density altitude, anything over, say, 5,000 feet, takeoff with a full-rich mixture may result in spark plug fouling, reduced engine power, or a rough running engine. In such a case the pilot should "stand on the brakes" and run the engine up to full power. He should then lean the mixture until the EGT shows approximately 100 degrees or so on the rich side of peak. If no EGT is available, he should lean until the engine starts to run rough and then enrichen until it runs smooth again and a little more. We do want a mixture which is richer than the one used in cruise, but full rich is not advisable when operating from a high-altitude airport. Once what feels like the right mixture has been achieved, the throttle can be reduced to idle until it is time to start the actual takeoff run.

Takeoff and Climbout

We're now ready for takeoff. We'll call the tower and, assuming no conflicting traffic, we are cleared for takeoff. At this point, while it is not actually required, it is generally good practice for the pilot to tell the tower the type of departure the pilot is intending to execute, such as: "Request left turnout, right turnout, downwind or straight-out departure." The tower will nearly always come back with "approved," but at least he now knows what we're going to do, and he can handle other traffic accordingly.

By the way, a request for takeoff clearance may be followed by either one of three types of instructions from the tower: "Cleared for takeoff," or "Taxi into position and hold," or "Hold short.' *Cleared for takeoff* means that you should taxi onto the runway without undue delay and should start your takeoff run. *Taxi into position and hold* means that you should taxi onto the runway and line yourself up with the centerline, but you may not actually commence the takeoff until told to do so by the tower. Variations on this may be such phrases as "Taxi into position behind the departing Learjet," or "Taxi into position behind the Cherokee touching down," or some such. At no time will a tower controller ever use the word *takeoff* except in conjunction with *cleared for*. The reason for this is that it reduces the chances for a misunderstanding to a minimum. Occasionally "Cleared for takeoff" will be followed by "Caution wake turbulence the departing 707." The reason for this is that heavy jet aircraft, when operating at low speeds, create wake turbulence, actually a pair of counter-rotating horizontal tornadoes, which can be deadly to light aircraft. When taking off behind a jet, a light aircraft should always lift off at a point on the runway before the point at which the jet's wheels left the ground, and he should then stay above the flightpath of the jet and, as soon as possible, turn upwind and away from the jet's climb path. The reason is that wake turbulence is not created until the jet becomes airborne (it is in no way related to jet blast which is only dangerous in close proximity to the jet) and it remains at or below the level of the aircraft while tending to float downwind. Wake turbulence has been known to remain lethal for several minutes and long distances behind a slow-flying jet and should not be ignored or taken lightly (Fig. 7-2).

All right, we've been cleared for takeoff. We advance the throttle all the way and, as the speed increases, use whatever right rudder is necessary to keep the airplane going straight along the centerline of the runway. If there is a crosswind we may want to turn the control wheel slightly (not all the way) toward the direction from which the wind is blowing. As we approach liftoff speed we slowly and tentatively pull back on the control wheel until

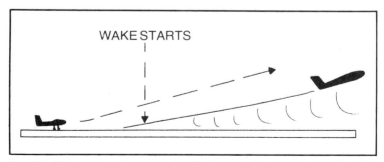

WAKE STARTS

Fig. 7-2. When taking off behind a jet, lift off before the point at which the jet's wheels left the ground and stay above the flightpath of the jet.

we sense that the airplane is developing a sufficient amount of lift to become airborne. As soon as the wheels have left the ground it is advisable to reduce the back pressure to let the airplane continue at a more or less level attitude until it has had a chance to build up more airspeed. Then, when the speed is safely above stall, we can again increase the back pressure and start our actual climb.

It seems appropriate here to talk briefly about *ground effect*. Ground effect is the added lift which is the result of the compression of the air between the wings (and, to some extent, tail) of the airplane and the ground. Because of this ground effect an airplane will appear to be ready to fly when, in fact, it is not. It will lift off and fly close to the ground, but if we then pull up without first gathering more speed, it will drop right down again. In flying, speed is the secret word and we *never* want to be caught with not enough of it, especially not when relatively close to the ground.

Before continuing with the climbout, let's look at two other-than-normal takeoff conditions. One is the short-field takeoff. Different pilots tend to advocate slightly different procedures in order to get off the ground in the shortest possible distance, and not every airplane responds in exactly the same manner as the next. It would not be advisable to attempt a short-field takeoff in an airplane with which the pilot is not completely familiar.

Here I shall describe the procedure which I have found to be most effective. The primary problem, of course, is to accelerate to flying speed just as quickly as possible without wasting any portion of whatever amount of runway is available. My favorite method is to stand on the brakes and to run the engine up to full power, with flaps up and the elevator held in a neutral or near-neutral position. The reason for this latter is that once we get rolling we want the airplane to be as clean as possible, presenting the absolute minimum amount of drag. Then, as soon as the engine has developed full power, I let go of the brakes and the airplane will accelerate more quickly than usual. As soon as something approaching liftoff speed has been achieved, I lower the flaps to their normal takeoff position and start, at the same time, to increase the amount of back pressure. If this has been done at just the right moment, the airplane will start to want to fly almost instantly. We can now actually pull it off the ground but then, assuming there is no obstacle to worry about, we'd better level off and let it gather speed before starting the actual climb. In extreme cases where there is, say, a fence at the end of the runway, it may be necessary to drop full flaps to literally jump over the fence. This works all right, but in some aircraft it is exceedingly difficult to gain adequate flying speed with full flaps and we may find ourselves mushing along unable to gain speed, simply neither climbing nor losing altitude. This is especially true in aircraft such as the Cessna 150 or 152 with those huge flaps and rather limited power. If that should happen—the condition is described as "being behind the power curve"—don't get impatient and don't try to make course corrections with anything but the shallowest of banks. Hold the nose down as much as possible without causing the aircraft to actually start to descend and slowly, *ever* so slowly, start to retract the flaps, just a tiny bit at a time. Gradually this will result in small increases in airspeed and eventually you'll be able to work yourself out of that box.

Another situation calling for special treatment is the so-called rough-field takeoff, usually involving grass strips. Here, too, our intention is to get the airplane off the ground as quickly as possible, but this time we must be concerned with preventing any damage to the nosewheel as a result of the rough ground. I start off the same way, holding the brakes and feeding in full power. But this time the flaps are in the takeoff position and the control wheel is held back all the way in my lap. Now, as soon as I let go of the brakes and the airplane starts to accelerate, the flaps will help to reduce the overall weight of the airplane on the wheels, and the up-position of the elevator will tend to lift the nosewheel off the ground long before the airplane is actually ready to fly. Now comes the tricky part. At this very high angle of attack, the airplane will want to lift off as soon as the speed is just a fraction above stall, actually a bit sooner because of the ground effect. But if we tried to pull it off, it would probably settle down again. The moment it begins to feel light, start releasing some of the back pressure, and just as soon as you feel that the wheels have left the ground, push the nose down and keep it down, flying along just a foot or so above the ground, until a safe climb speed has been reached.

One word of caution: Don't *ever* try to run the engine up to full power while holding the brakes, unless you've first made sure that there are no loose objects, gravel, twigs or similar debris, under or directly in front of the airplane. The prop will tend to pull such loose matter into itself and serious blade damage can easily result. As a matter of fact, when taking off from a gravel runway, the throttle should be advanced quite slowly to give the airplane a chance to start moving. Once it is moving at a reasonable speed, the chance of propeller damage is reduced considerably because the loose matter will be thrown behind the blades. For this reason, it is virtually impossible to make a successful short-field takeoff from a gravel strip without taking the chance of subsequently being faced with a healthy bill for propeller repair.

All right, we're now ready to climb. If the airplane in question is equipped with a fixed-pitch prop, we'll continue at full throttle. If it has a constant-speed propeller, we'll probably reduce manifold pressure and rpm to whatever setting is recommended by the manufacturer. The next thing is to get rid of the flaps, and, if the airplane has a retractable gear, this should have been pulled up as soon as we passed over the far end of the runway. As a general rule, it makes sense to leave the gear down as long as there is enough runway ahead that a landing could be accomplished straight ahead in the event of some malfunction. As soon as that point has been passed, the gear should be retracted. *Caution:* In some airplanes, particularly the retractable high-wing Cessnas (Fig. 7-3), retracting the gear activates all kinds of gyrations involving not only the gear itself but also all those gear doors. This will briefly increase drag by a quite considerable amount and in such aircraft gear retraction should not be initiated until adequate speed has been reached. Premature gear retraction in such aircraft could result in a loss of some altitude and worse, if the pilot's reactions are to prevent that loss of altitude by increasing back pressure, a stall could develop.

Retracting the flaps produces a reduction in the amount of lift being generated by the wings, and there might be a momentary feeling that we're losing altitude. For this reason, flaps should not be retracted until the airplane is at least 100 feet above the terrain at a speed which is safely above stall. Some aircraft react to changes in flap position by wanting to either raise or drop their nose, while others seem to show no such reaction at all. If a change in aircraft attitude results it should be corrected by first using the appropriate amount of forward or back pressure on the wheel and by following that with the necessary trim adjustment.

Throughout the takeoff run and climb we'll find that we have to hold right rudder in order to keep the airplane moving straight ahead. This is the result of something which is (rather inaccurately) referred to as *torque*. If the aircraft is equipped with rudder trim, we can use this rudder trim to eliminate the need for holding right rudder. But this should only be done after the desired climb has been established and the elevator trim has been adjusted accord-

Fig. 7-3. Retracting the gear on high-wing Cessnas results briefly in a considerable increase in drag (courtesy of Cessna).

ingly. Using rudder trim too early will only result in the need for repeated adjustments.

There are three basic types of climb angle which we may wish to choose. Climbing at or near the best angle of climb will result in a very steep flight path with the nose of the airplane way up in the sky, and forward visibility reduced to close to nil. Climbing that steeply is fairly uncomfortable for the pilot as well as the passengers, and is recommended only if there is a good reason for wanting to gain maximum altitude while covering a minimum of horizontal distance, such as trying to get over a cloud or a mountain. Climbing at the best angle puts the airplane relatively close to stall and any sudden increases in back pressure should be avoided. Also this is not a good time to bank steeply in one direction or the other, as any degree of bank drastically increases the stall speed.

Then there is the best rate of climb. This is the climb gradient which produces maximum gain in altitude over a given period of time (rather than distance). It is somewhat shallower than the best angle, and therefore is more comfortable for pilot and passengers alike. But it, too, is sufficiently steep to limit forward visibility in many airplanes, and more often than not pilots will opt for the third type.

This third type is referred to as cruise climb and is any climb gradient which is shallower than the best rate. To cruise climb simply means that we cover much more horizontal distance while gradually getting up to our selected cruise altitude. In addition to being more comfortable and to giving us at least the illusion that we're accomplishing two things at once—moving along horizontally while gaining speed—it is also somewhat kinder to the engine because of the increased cooling resulting from more efficient airflow.

Pilots have always and will continue to argue about which type of climb is the most fuel efficient when looking at the flight as a whole. Some insist that cruise climbing saves fuel and time, while others believe that a steep climb is the better deal. In actual fact there is no appreciable difference. You can try it in practice or you can use your trusty little electronic computer/calculator at home. Either way you'll find that assuming all other conditions to be equal there is no difference worth talking about. I know that this won't convince everyone, but it is fact just the same.

Cruise and Navigation

In the chapter on fuel consumption we will examine the importance of mixture control, the relative setting of rpm versus manifold pressure, and the selection of the right altitude to minimize the negative or maximize the positive effects of wind components. We will therefore ignore this subject here and concentrate on navigation and the correct use of the aircraft control surfaces.

There are three basic ways in which we can navigate and most of us use a combination of all three during the average flight. One, the oldest, is known as *pilotage*. Pilotage means that we look at the ground and determine our position and adjust our direction of flight by reference to visible landmarks. These days we rarely depend on pilotage to any meaningful degree, except when, for one reason or another, we are forced to fly too low to be able to receive signals from the available ground-based electronic navigation aids. But pilotage is most difficult when attempted close to the ground. Landmarks tend to be unreliable unless viewed in relation to other landmarks and then compared to what is shown on our Sectional charts (WACs are practically useless for this purpose). Hundreds of roads look like hundreds of other roads, and one small town looks pretty much like dozens of others from the air. As a general rule we should never assume that a given landmark is actually what we think it is unless we are in a position to relate it to at least *two* other landmarks. If three such landmarks are seen from the airplane and do appear to be positioned *exactly* the way they are shown on the Sectional, then and *only* then can we be reasonably sure that what we're looking at is what we think we're looking at. But when at 1,000 feet or less above the terrain, there is not an awful lot of country to be seen and this makes finding three such related landmarks difficult indeed. In addition, except in airplanes equipped with a wing leveler or autopilot, studying the charts while at the same time holding the airplane straight and level, can prove to be a most frustrating chore with the ground below being so close.

The second and most unreliable method of navigation is referred to as *dead reckoning*. The term has nothing to do with dying. It is, in fact, a bastardization of *deduced reckoning* and is therefore sometimes also spelled *ded reckoning*. It refers to simply taking up a heading which the pilot assumes will, under the known wind conditions, get him to his destination. He then simply holds that heading for as long as he believes it should take him to get there. And that's it. But the trouble is that there is no such thing as knowing the direction and velocity of wind with any degree of certainty for any length of time. Thus, simply holding a given heading for more than a few minutes is not likely to produce the desired course.

The last (and these days most popular) means of navigation is to fly by reference to VORs and other ground-based electronic navigation aids. This, of course, requires the appropriate instrumentation in the airplane in order for the pilot to be able to receive and interpret the signals provided by these nav aids. How this works is too complicated to be explained in a few words, and it is therefore the subject of a separate chapter.

Some Myths Dispelled

No matter the type of navigation to be used, it is the pilot's first and foremost responsibility to *maintain control of the airplane*, no matter the circumstances. This is something that pilots often tend to forget when things get a bit sticky or when something unforeseen occurs. A sudden surprise causes the pilot, whether VFR or IFR, to become preoccupied with some peripheral problem and he ignores the most *important* task, *controlling the airplane*.

This, therefore may be an appropriate time and place to once more try to dispell some of those stubbornly lingering misconceptions which pilots, especially low-time pilots, seem somehow loath to let go of.

The first and worst of these again has to do with that lingering belief that there is some relationship between an automobile and an airplane. In the automobile, the throttle controls speed. In an airplane, except while taxiing, it does not. In an automobile the steering wheel is purely for directional control. In an airplane, the control wheel is used for directional control in conjunction with the rudder, but it also controls speed.

Let's first talk about altitude control and the relation to it played by the elevator and the throttle. To put it into a categorical statement, the elevator, despite its name, does *not* control altitude. It controls the angle of attack, *period*. But (and when we're talking about the aircraft controls there always seems to be a "but"), since a flat angle of attack produces lift only at high airspeed while a large angle of attack will produce lift even at relatively slow airspeeds, the elevator is, in fact, the primary speed control. The throttle, on the other hand, is strictly an altitude control and has absolutely nothing to do with speed (except while taxiing).

I can already see all kinds of readers saying that I obviously have lost my marbles, so I better explain why what I have said is actually the case.

Let's say that we're flying straight and level at 55 percent of power, showing an airspeed (TAS or IAS, it makes no difference here) of 100 knots. We decide that that is kind of slow and we'd like to go a little faster. So we advance the throttle until the engine produces 65 percent of power. Logic would seem to dictate that this will make the airplane go faster. Except it does *no such thing*. Instead, it will cause the airplane to climb and, unless the pilot intervenes, it will set up a rate of climb at which that same 100-knot speed will be maintained. In order to make the airplane go faster (which was, after all, what we had in mind in the first place), the pilot will have to trim the nose of the airplane down. This will cause the elevator to move downward a fraction, thus reducing the angle of attack. Now, at this new angle of attack, the airplane will fly faster with the 65 percent of power. But if we now reduce the power setting back to 55 percent, instead of slowing up, the airplane will simply drop its nose and eventually settle on a rate of descent at which it will maintain the same speed at which it was flying straight and level at the higher throttle setting. This may seem confusing at first, but it is not if we finally get it through our heads that the *angle of attack is what flying is all about*.

An airplane trimmed for a given angle of attack will maintain a constant airspeed as long as that angle of attack is maintained. If more power is added, it will simply climb to maintain the same angle of attack and, in turn, the same airspeed. If power is reduced, it will drop its nose in order to achieve the same result. It is physically impossible for an airplane to change airspeed without a simultaneous change in the angle of attack. Thus, since the elevator and/or the elevator trim controls the angle of attack, it controls airspeed. The throttle, on the other hand, controlling the percentage of power produced by the engine and thus the amount of thrust created by the propeller, controls the position of the nose of

104

the airplane relative to the horizon—in other words, the *pitch attitude* or, quite simply, the rate of climb or descent.

I know of old-time pilots with thousands of hours in their dog-eared logbooks who still think of the throttle as the speed control and the elevator as the altitude control. They are dead wrong, but having become accustomed to using the controls the way they need to be used through sheer experience, they do just fine despite the fact that their thinking is based on incorrect assumptions. In that case you may say that obviously how one looks at these controls doesn't make any difference. The fact is that it *does*. It does when we find ourselves at the ragged edge of the flight envelope.

Here is an example: We've inadvertantly stayed too long on base as the result of a tailwind and now must make a very steep turn in order to be able to reestablish ourselves on a final which is more or less in line with the runway. In the middle of that steep turn we realize that we're losing altitude at a faster rate than we would like, so we pull back on the wheel in order to stop the descent. What happens? What will happen is that we'll reduce the airspeed. If this takes place at an indicated airspeed which is safely above stall, the decrease in airspeed resulting from an increase in the angle of attack will actually cause the rate of descent (or sink) to lessen. But if we *did* pull back on the control wheel when already within a mile or so of the aircraft's stall speed (at that particular angle of bank), the airplane will stall and possibly spin into the ground.

What we should have done in that situation was to either add power which would have produced a combination of angle of attack and thrust which would have permitted the airplane to fly on without losing altitude. Or, if no additional power was available, we should have *lowered* the nose. Granted, when we're close to the ground and are already losing altitude, the idea of lowering the nose is anathema, but it simply is the *only* way out. Lowering the nose will lessen the angle of attack, causing an increase in speed which, in turn, will stop the airplane from descending further. The only other alternative in this example would be to roll out of the turn and fly straight and level until things are back under control, in which case we'd probably have to start the approach to that landing all over again.

A similar situation frequently occurs on final approach. The airplane has been slowed up to the usual approach speed and we suddenly begin to realize that either because of miscalculations on our part or as the result of unexpected downdrafts, we're about to undershoot. The psychologically natural thing to do is to pull back on the control wheel in order to flatten out that glide and to make sure that we reach the runway. But that's not what's going to happen. What will happen is that the airplane will raise its nose. This, in turn, will cause a reduction in airspeed and conversely an increase in the rate of sink. Instead of having stretched the glide, we most probably have shortened it. Here the only way out would have been an increase in the throttle setting, because it is the throttle which controls the rate of descent (or climb) assuming no change in the angle of attack. Lowering the nose would have added airspeed but could probably not have achieved the desired result without the addition of power. (Always remember, *you can't stretch a glide*, but more about that later.)

A truly extreme example of all this is the *spin*. In a spin the airplane is in a near-vertical attitude rotating around its own axis and descending toward the ground at a moderate rate. The airplane is at an angle of attack which is simply too great to permit the generation of any lift at all. The only way to get out of such a spin is to push the nose *down*, an action which goes against everything our instinct demands. But pushing the control wheel forward will decrease the angle of attack to a point at which the airfoils are again able to generate lift. By doing that and, at the same time, applying hard rudder opposite to the direction of rotation of the aircraft, the aircraft will level off and start again to fly in a normal manner.

There is always a lot of argument about whether spin training should again be made a mandatory portion of regular flight training. Personally I am against it. It seems an unnecessary exposure of student and instructor to a very real accident potential. What *should* be taught (and rarely is) is that all important concept of the angle of attack. Any pilot who has successfully cleared his head of all those misconceptions about the real effects of aircraft controls on flight will learn to ignore his instincts and to do the right thing without having to think about it. This is difficult, I know, because our instincts are based on living and functioning on firm ground and not up in the air. Flying is not something that comes naturally to the human being. It has to be *learned* and, in the process, certain instinctive reactions have to be *unlearned*.

Function of the Rudder

So far we have talked, at considerable length to be sure, about the control of the airplane with reference to its pitch, meaning its movement in the vertical sense. Now let's look at directional control. Here let's get one thing straight right from the beginning. *The rudder is not a turn control.* The purpose of the rudder is to control yaw. In plain English this means that the purpose of the rudder is to keep the tail of the airplane behind its nose. This may sound silly since the airplane is a rigid contraption on which the rudder is always behind the nose. But is it? With respect to the construction of the airplane, true enough: it always is. But with respect to the direction with which the airplane flies through the air, it frequently is *not*.

Again we need to resort to some examples in order to explain what this is all about. Let's assume we're flying straight and level. In this case, the tail—and with it the rudder—is in a straight line with the nose of the airplane, moving through the air directly behind it (Fig. 7-4). Now, let's push hard right rudder. The nose of

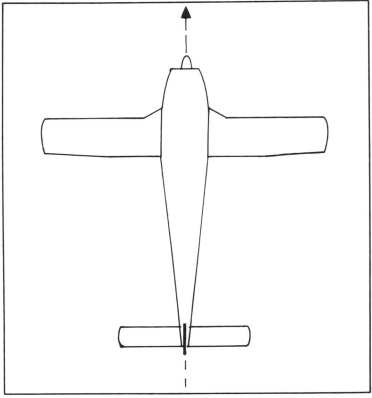

Fig. 7-4. With the rudder held straight, the nose of the aircraft points in the direction of flight.

the airplane will veer to the right and the tail assembly including, of course, the rudder, will veer to the left (looking at it in the direction of flight). But will the airplane turn? No, not really, not if the wings are held in a truly horizontal attitude. It will actually continue in the direction in which it was flying before, but it will do so slightly sideways (Fig. 7-5). This is referred to as a *skid*, and

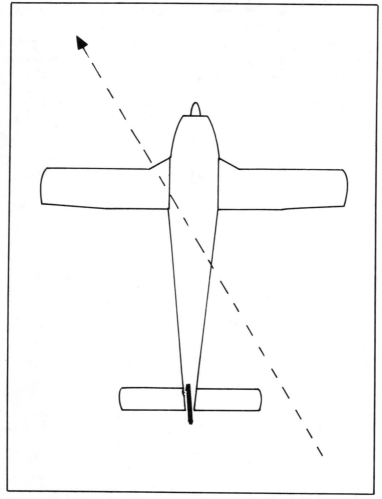

Fig. 7-5. When the rudder is deflected in one direction or the other while the wings are kept level, the aircraft will continue to fly straight ahead, but the nose will point toward the direction of the rudder deflection. The aircraft is referred to as skidding.

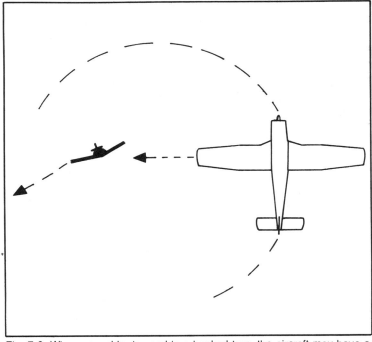

Fig. 7-6. When no rudder is used in a banked turn, the aircraft may have a tendency to slip toward the center of the turn.

even slight skids can be seen by the pilot in the cockpit by watching the action of the ball in the glass tube at the bottom of the turn-and-bank indicator. In the above example, the ball will show up in the extreme left of the glass tube, indicating that the pilot must apply left rudder (or reduce right rudder) in order to reestablish coordinated flight.

But if the airplane keeps its tail straight behind its nose automatically, then what's the use of having a rudder in the first place? The fact is that the airplane does not necessarily keep its tail straight behind its nose, and that is where the rudder comes in. When we turn the control wheel to one side or the other, thus activating the aileron control, one wing will rise while the other one will drop. The action results in a combination of two reactions by the airplane. One is that the airplane will start to turn in the direction of the lower wing. The other is that is will begin to slide sideways toward the lower wing as a result of the gravity pull of the earth (Fig. 7-6). When that happens, the nose of the airplane will actually point away from the direction of the turn, while the tail slides sideways toward the direction of the turn. This is called a *slip* (Fig. 7-7). Again the airplane is flying more or less sideways, and now we need the rudder in order to push it over so that it repositions itself in a straight line behind the nose, relative to the direction in which the airplane moves through the air.

Here things tend to get a bit complicated. The logical thing would be to assume that such a turn will always require the use of rudder on the side of the turn. Sometimes this is true and sometimes it is not. The difference has to do with airspeed, the degree of bank, and, to some extent, that famous phenomenon known as torque. Because of the multitude of interactions of all these factors, there is no simple way to say that any such-and-such turn will require such-and-such rudder. It is something which has to be experienced, and then learned through experience. Suffice it to say that in a turn, depending on the airspeed, the degree of bank, the amount of power produced by the engine and the resulting "torque," the tail of the aircraft may either be on the side of the turn

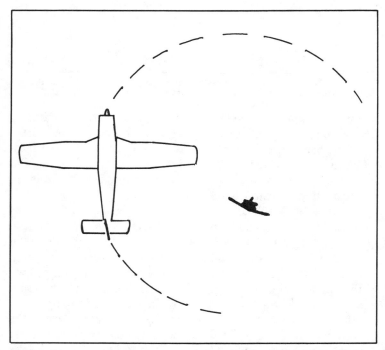

Fig. 7-7. When the aircraft tries to slip toward the center of the turn with the nose pointing away from the center of the turn, inside rudder—rudder deflected toward the center of the turn—is indicated.

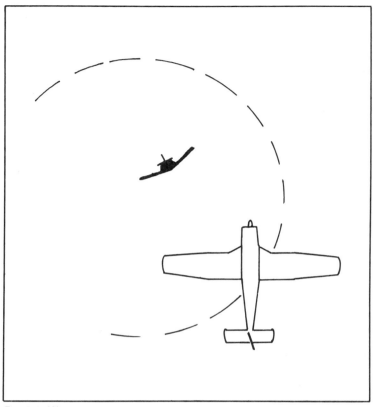

Fig. 7-8. When, while slipping toward the center of the turn, the nose of the aircraft turns too far toward that center, opposite rudder is indicated.

or it may want to move to the outside of the turn. In the first instance, a certain amount of rudder pressure is needed on the side of the turn in order to keep the ball centered and, of course, the tail in line with the nose. In the second instance, a degree of rudder pressure in the direction *away* from the turn must be employed to achieve the same result (Fig. 7-8).

Now it is quite true that turns can be made without using the rudder, and as long as these turns are sufficiently shallow, it really doesn't make much difference if they result in a slight degree of skidding. In cruise, many of us will fly for hours with our feet on the floor, making minor course corrections without ever touching the rudder. There is nothing wrong with that. But when it comes to *steep* turns, especially at low speeds, complete coordination is important and the rudder becomes an important control.

One more comment with reference to turns. While the ailerons are needed to *initiate* a turn, they must be neutralized and sometimes even moved to the opposite direction in order to maintain a constant rate of bank. Otherwise the bank will tend to continue to increase, eventually resulting in a screaming spiral.

Altitude

Now, having set that subject to rest, let's look at the effect of altitude on the performance of an airplane. The only real difference between high altitude and low altitude is that at high altitudes the air weighs less per volume than it does at lower levels (and that, in case of extremely high altitudes, there isn't sufficient oxygen available for the human body to continue to function properly). But the weight of the air, the fact that it is what is referred to as *thin*, has both positive and negative effects on the way the airplane flies and on the amount of power which an engine is capable of producing.

To put it simply, thin air results in a reduction of lift, but it also results in a reduction in drag. The action of the propeller is less efficient because there is less air for it to bite into. There is a barely noticeable reduction in the effectiveness of the control surfaces, assuming no difference in true (not indicated) airspeed between high and low altitudes. And the engine, requiring a given amount of air along with its fuel, produces less power.

In practice we rarely notice the reduction in lift because, as a general rule, we tend to fly at a higher true airspeed up high than we do down low. But the fact is that an airplane flying through thin air requires a greater angle of attack than one flying at the same true airspeed through the denser air available at lower flight levels. It does become noticeable at extremely high altitudes, such as above 18,000 feet for the average normally aspirated single-engine airplane. Here, despite the fact that the true airspeed may be considerable, it tends to assume a nose-high mushing attitude in order to maintain level flight.

The reduction in propeller efficiency varies with the type of propeller. For aircraft equipped with fixed-pitch props it is quite considerable. In aircraft with constant-speed props it is less so, especially if a relatively low rpm setting is used. At such a setting the propeller is able to take huge bites out of the air with each revolution, thus, to some degree, reducing the negative effect of the lessened air density. (This is also one of the main reasons why aircraft with fixed-pitch props are rarely if ever turbocharged. The high-altitude advantage produced by the turbocharger would be negated by the inefficiency of the propeller.)

The loss of control-surface effectiveness is another phenomenon which the pilot is not likely to be aware of. As a matter of fact it may *seem* as if they have become *more* effective, resulting in a certain skittishness. Actually, even though it may seem contradictory, both are true. The reduction in control-surface effectiveness results only if the aircraft is moving through the air at the same (true) airspeed as when it moves through denser air. Then, because less air strikes the control surfaces at the same velocity, increased control-surface deflection is required in order to get a comparable result. But since we nearly always move through the high-altitude air at much greater speeds this effect is negated by the increase in speed. And when we get really high up a certain skittishness in response to control inputs becomes apparent, making it difficult to trim the airplane for continued level flight.

And then there is the engine. In normally aspirated engines the intake-manifold pressure is directly related to and limited by the ambient atmospheric pressure. It follows therefore that since the atmospheric pressure at altitude is markedly lower than at sea level, less manifold pressure is available. But the relationship between the amount of air and fuel which must be available for efficient combustion *doesn't* change. As a result, with less air pressure in the manifold, less fuel can be used and, in turn, the engine produces a reduced percentage of power.

In actual operation, the combination of the various negative results of lessened air density are more than overcome by a quite considerable reduction in drag, which is why we actually move faster through the air despite a lesser percent of power and a lower indicated airspeed. This is what makes flying at high altitudes more economical than flying down low, as is explained in greater detail in the chapter on fuel consumption. Also explained in that chapter is the effect of wind which, more often than not, blows at vastly greater velocities the higher we get. (I once had a 150-knot tailwind at 15,000 feet in an airplane with a TAS at that altitude of 115 knots. If I'd been flying in the opposite direction I would have been moving backwards across the ground.)

Wind

Now let's climb down again from those high altitudes and look at the effect of wind on navigation which is easier to visualize when we're relatively close to the ground. Wind is, in fact, a moving mass of air, and the airplane operates within that mass and relative to it and not to the ground. The best way to visualize that is to think of the airplane as being hermetically sealed within an invisible "envelope" of air (Fig. 7-9). If that envelope moves in one direction or another, the airplane will inexorably move with it. As a result, if the wind blows from the right, the airplane will move sideways across the ground to the left, while at the same time moving straight forward within its envelope of air (Fig. 7-10). What results is a diagonal flight path across the ground. The inexperienced pilot, seeing the airplane moving across the ground in such a crabbing

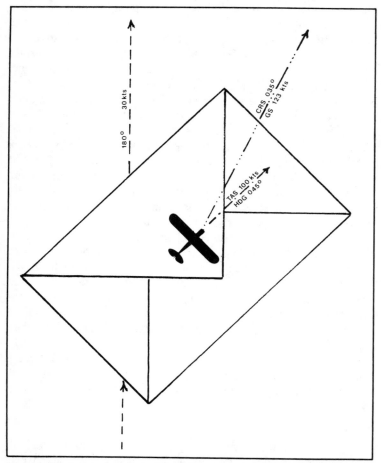

Fig. 7-9. An aircraft must be thought of as flying within an envelope of air. The course over the ground is the result of the heading of the aircraft within that envelope of air, and the movement of that envelope of air as the result of wind.

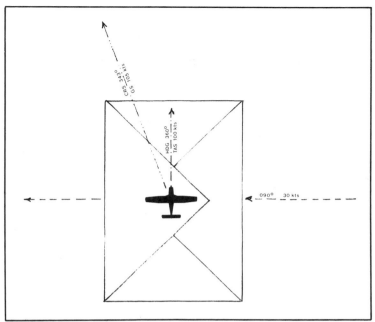

Fig. 7-10. A 90-degree crosswind, when no heading correction is made, will result in considerable downwind drift and a slight increase in ground speed. If the appropriate wind correction is made, the nose of the aircraft will point upwind and the true airspeed will be higher than the ground speed.

motion, will automatically think of it as a yaw (even though he may not be aware of that term) and he may be tempted to use his rudder to get the nose aligned with his direction of flight relative to the ground. While this may feel like the right thing to do, it is totally *wrong*. The only way he can align the nose of his airplane with his direction of flight (again, relative to the ground) is to fly directly into or straight away from the wind, which, quite obviously, is not what he had in mind.

The airplane doesn't know where the ground is and it couldn't care less. Flight is relative to the surrounding air mass, and the fact that the airplane *appears* to be flying at an awkward angle is strictly an illusion created by looking at the ground. The speed and direction of aircraft movement is the result of the combination of the thrust produced by the propeller minus induced and parasite drag and plus or minus the direction and velocity of the wind (Fig. 7-11).

There are times when this statement must be modified to a degree. When an airplane passes through a wind shear, an area in which wind changes either direction or velocity, the airplane will not react to the change instantly. The reason is something which we refer to as *inertia*. Inertia means that a moving body will want to continue to move in the same direction and at the same speed as it has been moving, or a non-moving body will resist movement. To use a rather extreme example: An airplane is descending at a TAS of 100 knots into a 30-knot headwind, producing a ground speed of 70 knots. Suddenly the wind shifts 180 degrees to a 30-knot tailwind. For a brief moment the airplane will be moving through the air at only 40 knots because there is a 60-knot difference in the wind velocity relative to the direction of movement of the aircraft. The airplane would stall. This phenomenon doesn't last long. The airplane will quickly adjust to the effects of the new direction and velocity of the airflow and, in the above example, if it didn't stall it would quite quickly regain that 100-knot TAS which would then have increased the ground speed to 130 knots. While this example is more extreme than is likely to be encountered, fairly drastic windshear situations often do occur close to the ground. Therefore, especially when gusty winds are forecast or being experienced, a pilot would be well advised to maintain a larger than normal margin over stall speed until it is time for the flare and landing is assured.

Landing

Now let's talk about landings. Every flight involves at least one landing and landings are practiced more diligently and with more frequency than any other maneuver. One fact ought to be stated right at the beginning: *No two landings are ever exactly alike.* Thousand-hour pilots occasionally make lousy landings and a student may now and again "grease one on." Landings are affected by pilot judgement, aircraft performance, wind, and the condition and length of the runway. Because at least some of these conditions may vary with each landing, it is my personal belief that a dozen touch-and-gos on a Sunday afternoon at the home airport are a waste of time and fuel. Making one landing after another at the same airport under the identical conditions only teaches how to make an acceptable landing at that particular airport under those specific conditions. It *in no way* prepares the pilot for landing some other day at a different airport under other conditions.

It has been said many times and it is no less true that a good landing is the culmination of flying a good approach. The approach is all important and deserves to be looked at in great detail.

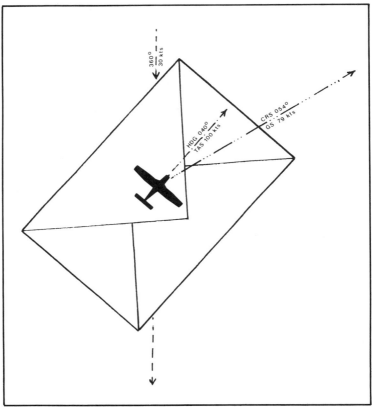

Fig. 7-11. When a headwind causes the envelope of air to move in a direction opposite to that in which the aircraft is flying, a considerable reduction in ground speed will result.

First, there is the pattern. There is a good reason why students are urged to always fly a complete pattern as part of the approach to a landing. The pattern consists of a *downwind leg*, a *base leg*, and the *final approach*. Most patterns are flown to the left (every turn being a left turn), but some airports have right-hand patterns and both need to be practiced, because from the point of view of the pilot they are *not* identical.

Downwind: The accepted routine for entering the downwind leg, whether at a controlled or an uncontrolled airport, is to approach it at a 45-degree angle which intercepts the downwind leg at some short distance downwind from the center of the runway or the tower. By the time the aircraft then turns downwind, it should be at an altitude of somewhere between 800 and 1,000 feet above

117

the runway elevation and it should have been slowed down to the approach speed applicable to the type of airplane being flown, generally about 50 percent above the stalling speed of that aircraft. Thus, by the time the airplane is opposite (abeam) the center of the runway (or, in the case of a controlled airport, the tower), it should be at the above mentioned pattern altitude and it should have been trimmed to fly straight and level at a speed approximately 50 percent above stall. (If the aircraft has retractable gear, the gear should now be down.) Partial flaps may be deployed at this point, or the pilot may prefer to delay using flaps until he is established on final. That is a matter of personal preference and of the type of airplane being flown (and, in the case of dual flight instruction, of the way the instructor wants it done). It is vitally important that the airplane has been trimmed right. Simply maintaining level flight by holding back pressure is not enough, as this will greatly increase the pilot's workload during the balance of the approach. The downwind leg is now flown parallel to the runway until the approach end of the runway is approximately 45 degrees behind the aircraft. During this time we must carefully examine the airspace in the vicinity of the airport to be sure to be aware of other traffic. If there is a control tower, the controller is likely to request that we follow a given aircraft ahead. In that case, detecting that aircraft is often difficult and if we do not spot it, we should so inform the tower who may then try to help us to find it.

In such a case, where we are asked to follow another aircraft, we should not be tempted to turn base at a point nearer the airport than the one at which the preceding aircraft made the turn, even if that takes us way beyond the point at which we would normally initiate base. Turning too early could result in our following the preceding aircraft too closely which may lead to the need to go around.

Base leg: The base leg is a flight path at right angles to the runway. It is the connecting portion between downwind leg and final approach, and it should usually be used to start losing a limited amount of altitude. Now we are fairly close to the actual landing, and in order to avoid crowding too much pilot workload into the last few moments, it might be advisable to lower the flaps some more. And remember, to cause the airplane to lose altitude, *don't* put the nose down. Altitude is controlled by the throttle, so reduce throttle and, if necessary, trim the nose *up* a fraction more. Just before the airplane reaches what might be thought of as an extension of the centerline of the runway, we now roll into the final approach which lines us up with that centerline of the runway.

Under conditions of strong crosswinds, the base leg can become a problem. If by turning from downwind leg to base we turn into the crosswind, then there is nothing much to worry about. It simply means that it will take a bit longer before we have to turn final. If, on the other hand, by turning base we turn away from the wind, then the possibility is considerable that we may be blown past the centerline of the runway before we've had an opportunity to get established on base. In extreme cases, it may be necessary to skip base altogether and to make one long turn all the way from the downwind leg to final. Under such a condition we may be tempted to bank more steeply than normal. *Don't!* Remember that the aircraft is flying at a slow speed and that stall speed increases *drastically* with banking. It's a lot better to abandon the approach and try again than to take a chance of stalling into the ground.

Final approach: As soon as we're established on final, we should call the tower and tell them so. Remember, at a controlled airport we cannot legally land unless we have received and understood the clearance: "Cessna Three Four Zulu, cleared to land." If the airport is uncontrolled, we've got to keep an eye on the airplane which has landed or is landing ahead of us to be sure that he is safely clear of the runway before we touch down.

If the aircraft engine is carburetor equipped, we should now apply full carburetor heat as a precaution against the possibility of carburetor ice which is greater when the engine is operating at idle or a low power setting. If the aircraft is equipped with a constant-speed prop, we now push control all the way forward to the high-rpm setting. While all this is going on, we also need to set up a glide path which will get us over the fence and onto the approach end of the runway at a speed which won't result in a long float when we finally have to flare.

Different pilots have different preferences with reference to the way the final approach is flown. Some, and this includes virtually all flight instructors, like a relatively steep glide, using full flaps and the engine at idle. Again, the pilot should trim the airplane so that it maintains a steady rate of descent at a steady slow speed without control input by the pilot. Then, if slight adjustments are called for, the required pressure on the control wheel is minimal. But always remember that if it looks as if the aircraft may undershoot, *don't* pull back on the control wheel. At this point the speed of the airplane is too slow to respond in the desired way. Use the throttle! Add some power and the airplane will respond by flattening out the glide path.

The advantage of this type of steep descent is that it reduces the danger which might result from sudden and unexpected downdrafts. The disadvantage is that a slight mishandling of the controls by the pilot can easily result in a sink rate which is too great to be rectified in time, especially by an inexperienced and, therefore, nervous pilot. Also, unless the flare is initiated at just the right moment, it can result in a very hard "arrival."

The other extreme is to use a low, flat approach, usually with only partial flaps, and always with power. In this case the airplane is trimmed nose-high for a slow-flight configuration, and the power is constantly adjusted to maintain the desired glide path to assure touchdown at the right spot. Here the pilot must be ready to add considerable power in the event of a sudden and unexpected downdraft because the distance between the aircraft and the ground is critical.

Most experienced pilots will use a combination of the two types of approaches described above. One way or the other, the right hand of the pilot should remain on the throttle throughout the final approach. This is important for two reasons. One is that it somewhat reduces the tendency of the inexperienced pilot to misuse the elevator controls by hauling back on the wheel in order to try and stretch the glide. The other is that it acts as a constant reminder that it is the throttle which must be manipulated in order to assure touchdown at the right spot, and not any other control.

Flare: The flare is that very last maneuver initiated when the wheels of the aircraft are within inches of the runway. It simply means that we apply full back pressure to cause the airplane to stall and thus to settle down on the runway.

But this sounds simpler than it actually is. First of all, the pilot can only guess at the distance between the wheels and the concrete; he can't see it and therefore never really knows for sure. Second, if the airplane is flying more than a mile or two above stall speed, instead of settling down it will simply float and keep on floating, causing the inexperienced pilot to want to push the nose down in order to get it on the ground. This will either result in a busted nosewheel or it will cause the airplane to hit the ground and then bounce back up into the air. The only way to get the airplane to stay on the ground is to slow it down to less than lift-producing speed. If that cannot be done before the better portion of the runway has been passed, go around and try again. One way to cause a floating airplane to sit down is to get rid of the flaps. Flaps increase lift and if they are suddenly eliminated the airplane will quite quickly settle to the ground—unless, of course, the excess speed is extreme.

I cannot emphasize enough the importance of always trimming and retrimming the airplane throughout each segment of the approach. Among other advantages, it greatly reduces the amount of muscle power needed to pull back hard enough on the control wheel to effect a flare. In some aircraft, if they are incorrectly trimmed, the amount of force needed to flare can be too great to be handled by the average person, resulting in a nosewheel-first touchdown which is hard on the airplane and is likely to result in a bounce and one of those embarrassing multiple arrivals.

Crosswind landings: Crosswinds produce literally an endless variety of problems and, since any wind other than one blowing straight down the runway is, in fact, a crosswind, we find that we have to deal with them more often than not. In the old days, crosswinds seemed to be less of a problem. Actually, those old taildragger-type airplanes were pretty nearly incapable of making a successful crosswind landing, and as a result practically every airport worth its name had runways criss-crossing one another so that, no matter the direction of the wind, there was always one which pointed right into the wind. No more. Tricycle-gear airplanes are supposed to be landable in all manner of reasonable crosswinds, and, since criss-crossing runways reduce the potential acceptance rate of an airport, they have been generally eliminated in favor of longer and/or parallel runways. This means that whenever we take off these days, we'd better be prepared to accept a crosswind landing by the time we get to where we're going. (Obviously, since any one runway can only point in two directions, that leaves 358 degrees of direction which are technically crosswinds of greater or minor dimensions.)

There are all kinds of crosswinds: steady crosswinds blowing at a constant velocity at a given angle to the runway; gusty crosswinds which blow hard and soft at unpredictable intervals; crosswinds which blow at and around obstructions, like hangars or other buildings, and which cause a wild mixture of seriously disturbed eddies on the leeside of those buildings. And occasionally there are crosswinds blowing in different directions over different portions of the runway. (Once, arriving at Palm Springs, California, I was informed by the tower that the wind at the western end of runway 12/30 was blowing from 280 degrees at 15, that in the center of the runway which is 7,000 feet long, the wind was out

121

of 030 degrees at 10, and that at the east end it was out of 150 degrees at 20. They added, "Runway your choice." Some choice!)

There are a number of basic precautions which a pilot should take when flying an approach to and landing at an airport under crosswind conditions. The first and probably most important is to use either a minimum or no flaps. Flaps tend to greatly increase the effect of the crosswind on the airplane; therefore, don't use them unless you happen to be flying an airplane which cannot be successfully flown sufficiently slowly without them to permit a landing. (Offhand I can't think of a single single-engine aircraft that cannot be landed without flaps.) Second, while on final, turn the control wheel slightly (not too much) toward the direction from which the wind is blowing. If the wind is strong enough, this will result in a slightly sideways crabbed approach which, if you handle the ailerons just right, will go straight down the extended centerline of the runway (Fig. 7-12). If that doesn't seem to work out right, you might use a bit of opposite rudder. This will result in a degree of uncoordinated flight, but under these circumstances there is nothing wrong with that; but you may have to increase power a bit, as such cross-controlling may cause an increase in sink. One way or the other, be sure not to raise the upwind wing. Raising it would expose the underside of the wing to the effect of the wind and could result in being blown completely off course.

When we arrive over the runway in such a crabbed condition, it's nothing to worry about. As soon as we flare (again without raising the upwind wing) we hit the downwind rudder hard. This will cause the nose to swerve into the direction of our movement across the ground and we'll find that we're landing straight ahead. But keep holding the ailerons into the wind. Even when the airplane is already on the ground, the wind can play havoc. Keeping the ailerons pointed toward the wind, meaning that the upwind aileron points upward from the wing while the downwind aileron points downward, will spoil the lifting capability of the upwind wing and thus reduce the potential effect of that wind on the aircraft.

Shutdown

Now we're back on the ground. What was said about taxiing in the beginning of this chapter still holds true now. Just remember that, once ready to shut things down, always turn off all radios before shutting down the engine. It helps to protect the radios from sudden current fluctuations. And then, once in the tiedown area, tie the airplane down *yourself*. Don't trust some line personnel.

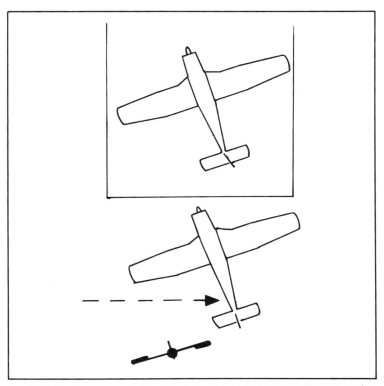

Fig. 7-12. During a crosswind landing, keep the upwind wing low throughout the approach and touchdown. The aircraft will appear to be approaching the runway at an angle. Just before touchdown use hard rudder in the direction away from the wind in order to straighten the nose.

Parked aircraft, especially high-wing aircraft, have been known to be pushed around by sudden gusts of wind, and it would be depressing to come out to the airport in the morning and find your bird sitting on top of the wing of a Learjet. And ask to have the tanks topped off right then. The walls of partially full fuel tanks tend to collect water vapor which then becomes water in the tank. The sooner the tanks are topped off, the less the chance of collecting a cupful of water when draining the sumps before the next flight.

Chapter 8
The Airspace

In order to fly today it is becoming increasingly important that we understand the true meaning of the designations by which the FAA describes the different portions of this airspace, and the rules which govern their use. Basically there are only two major categories, *controlled* and *uncontrolled*. But the amount of uncontrolled airspace is constantly shrinking, while at the same time, the controlled airspace is being portioned off into more and more individual sections, the degree of control over which varies.

The term *airspace*, as generally used in aviation, refers to the navigable airspace between ground level and 60,000 feet agl (Fig. 8-1). The controlled portion of this airspace is divided into *airport traffic areas, airport control zones, airways, control areas, control zones, continental control area, terminal control areas, transition areas,* and *positive control areas*. In addition there is a variety of special-use areas referred to as *warning areas, restricted areas, prohibited areas, jet training areas and military operations areas*. With a degree of difference in the manner in which all (or most) of these areas may be used by the average pilot, and in the mandatory equipment requirements it is only natural that things can get to be a bit confusing.

Uncontrolled Airspace

Uncontrolled airspace is that portion of the airspace over which air-traffic control has no jurisdiction. In uncontrolled airspace it is perfectly legal to fly VFR when the weather is above

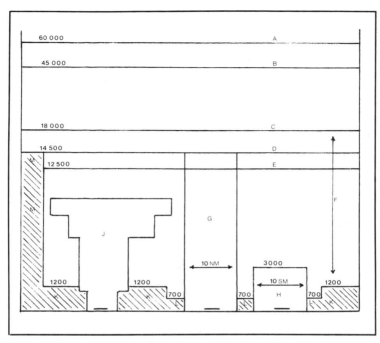

Fig. 8-1. The navigable airspace. All figures from 12,500 on up represent altitude in feet msl. All other figures represent altitude in feet agl. A: Airspeed beyond the jurisdiction of ATC; B: Controlled airspace above the top limits of the jet routes; C: Positive control airspace; D and including C: Continental control area; E: Encoding altimeters mandatory in controlled airspace; F: Airways; G: Airport control zone; H: Airport traffic area; J: Terminal control area; K: Uncontrolled airspace underlying the airways; L: Uncontrolled airspace underlying transition areas; M: Uncontrolled airspace outside of the areas covered by airways and other controlled airspace.

the minimums applicable to VFR flight in uncontrolled airspace, or to fly IFR, if weather conditions are below those minimums. In either case, there is no need to communicate with anyone or to obtain any type of clearance.

Uncontrolled airspace is all the airspace underlying controlled airspace, usually up to either 700 or 1,200 feet agl, except in the vicinity of controlled airports. Except where otherwise indicated on the charts, the airspace from to ground up to 700 feet agl below a transition area and up to 1,200 feet agl below airways and other controlled airspace not associated with an airport is uncontrolled. In areas in which there is no controlled airspace below 14,500 feet (of which there are very few left) the airspace is uncontrolled up to 14,500 feet which is the base of the continental control area which covers all of the United States except the extreme eastern portion of Alaska.

Controlled Airspace

Here we are faced with all those differing degrees of control and we'll have to take them one at a time. Some portions of controlled airspace exist only with reference to IFR operations, some are applicable to VFR aircraft only, and some affect both types of operation, in certain instances excluding VFR operations altogether.

Airport Traffic Areas

An airport traffic area is the area within a radius of five statute miles around an airport with a control tower during the hours when that control tower is manned and in operation (Fig. 8-2). It extends upward from the ground to (but not including) 3,000 feet agl. Airport traffic areas and the rules governing them apply to VFR aircraft only and do not exist for aircraft operating IFR. Any VFR aircraft operating within an airport traffic area must maintain two-way radio contact with the tower regardless of whether the operation involves a takeoff from or a landing at the controlled airport (or any other airport located within such an airport traffic area), or whether it is just passing through. Airport traffic areas are not shown on any aeronautical chart.

In practice this means that operations to or from controlled airports can be conducted only if the aircraft is equipped with at least one working two-way com radio. While en route, when overflying a controlled airport, VFR aircraft must be at 3,000 feet agl or higher, unless they are able or willing to communicate with the tower. If the pilot does not wish to climb to above 3,000 feet agl, the procedure involves tuning to the tower frequency and announcing: "ABC Tower, Cessna One Two Three Four Foxtrot, five east, request clearance to transit airport traffic area at two thousand, westbound, over." In most instances the clearance will be given along with a report on any traffic operating in the area, the current altimeter setting, and a request to inform the tower when the aircraft is again clear of the airport traffic area. In order to know when this happens, the pilot should try to figure out how long it will take him to fly 10 statute miles at his current ground speed (10 statute miles being the diameter of the airport traffic area at its widest point). Thus, if his ground speed is 140 mph, it will take him four minutes and 17 seconds.

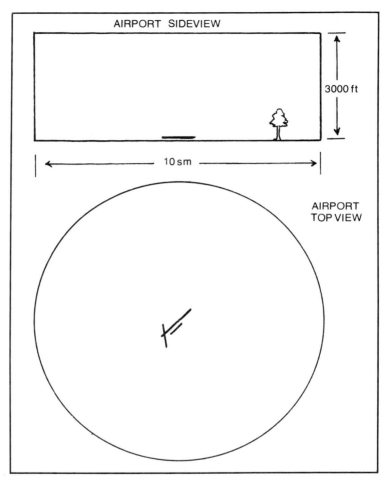

Fig. 8-2. Airport traffic area.

Airport Control Zones

Airport control zones are quite similar to airport traffic areas except that their purpose is different. While airport traffic areas apply only to VFR aircraft, airport control zones apply only to IFR aircraft and to aircraft operating under a Special VFR clearance. Like airport traffic areas, airport control zones are more or less circular and 10 miles in diameter, but their shapes vary somewhat and are shown on Sectional and WAC charts (Fig. 8-3). In many instances they include extensions to accommodate instrument departures and arrivals. Unlike the airport traffic areas they extend from the ground upward to the base of the continental control area (or, where not underlying the continental control area, they have no upper limit).

Airport control zones exist only in conjunction with airports with contol towers during the hours when the tower is manned and in operation, or in conjunction with airports where a flight service station provides the applicable services for IFR pilots. When operating within a control zone, all IFR aircraft are at all times under the control of the tower or flight service station, and are required to maintain two-way radio communication. Since under normal circumstances IFR aircraft are under ATC control throughout the entire flight there is no particular need to pay attention to the time at which the aircraft enters the control zone. It is only in the rare instance when an IFR aircraft approaches a control zone from uncontrolled airspace that it becomes important.

Airways

The airway structure consists of two levels, the low-altitude so-called Victor Airways (V-000) which are in effect at altitudes below 18,000 feet msl, and the Jet Routes (J-000) which have been established between 18,000 and 45,000 feet. For the purposes of this book we are concerned only with the Victor Airways. The reason for the term *Victor* (the phonetic alphabet designation for the letter V) is that they are referenced to VORs, VOR/DMEs and VORTACs, more often than not representing a straight line between two such VHF nav aids.

All Victor Airways are, in fact, controlled airspace, with the control extending from 1,200 feet agl (except where otherwise indicated on the charts) up to but not including 18,000 feet. They are indicated on aeronautical charts by the centerline of the airway and extend to a distance of four nautical miles to either side of that centerline. The fact that airways are controlled airspace is of no interest to the VFR pilot. He has every right to operate along the airways, across them, or to otherwise intercept them in any way he desires without any requirement of having to be in contact with air-traffic control or any other FAA facility. As far as he is concerned, they exist only in the navigational sense and in no other.

Things are different for the pilot operating IFR. For him, the term controlled airspace does exist, and he is required to maintain contact with ATC and to adhere to ATC clearances unless or until he decides to cancel his IFR flight plan.

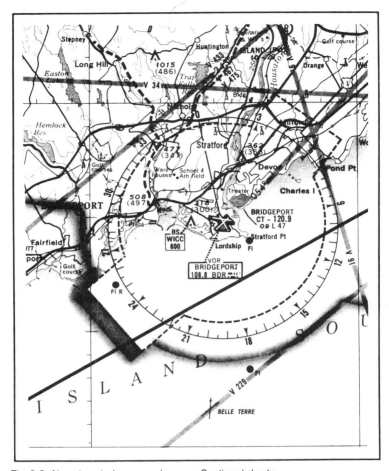

Fig. 8-3. Airport control zone as shown on Sectional charts.

Each airway or section of an airway involves certain minimum altitudes which are shown on the Jeppesen and Government low-altitude IFR charts, but not on the Sectionals and WACs (Fig. 8-4). They are the minimum en-route altitude (MEA), which is the minimum altitude between two fixes which assures acceptable navigational radio reception and also meets the necessary obstruction-clearance requirements; the minimum reception altitude (MRA) which is the lowest altitude required to receive reliable signals from nav aids needed in a particular route segment; and the minimum obstruction-clearance altitude (MOCA) which is the lowest possible altitude between fixes which meets obstacle-

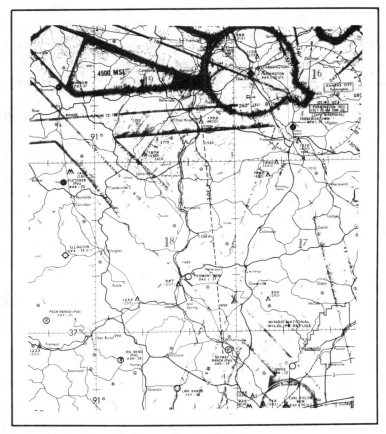

Fig. 8-4. Victor airways as shown on Sectional charts.

clearance requirements for a given segment. At this altitude, radio signal reception is guaranteed for only 22 nautical miles from the VOR. While these minimum altitudes also only apply to IFR aircraft, VFR pilots would do well to be aware of them as, especially under conditions of haze or other obstructions to visibility, adherence to them will assure a safe flight.

Since 1975 an additional restriction has been imposed on aircraft operating on airways (or in other controlled airspace) at any altitude above 12,500 feet except within 2,500 feet of the terrain in mountainous areas. Here aircraft are required to be equipped with a transponder and an encoding altimeter in order to operate legally. Aircraft not so equipped can request permission to climb to above 12,500 feet in order to get over the tops of some overcast or for similar reasons, and such requests will usually be granted. (Or they can, of course, simply cancel IFR and climb up there VFR, because there is no way for anyone on the ground to determine at what altitude they happen to be operating.)

Transition Areas

Transition areas are airway extensions which are established to permit a safe transition from the en-route altitude to the instrument-approach procedure. They are shown on the Sectional and WAC charts by a shaded pinkish outline (Fig. 8-5). The base of such transition areas is usually 700 feet agl unless otherwise indicated on the charts. Transition areas, like the airways, are controlled airspace, meaning that VFR pilots need pay no attention to them, while IFR pilots must maintain contact with the appropriate ATC facility, usually approach control.

Control Areas

The term *control area* applies to all airspace, including but not limited to airways and transition areas, over which ATC exercises control of IFR traffic. The VFR pilot can safely and legally ignore all control areas (but not positive control areas), while the IFR pilot must maintain communication with ATC. This can get to be a bit confusing, because there are, for instances, terminal control areas (TCA) which are, in fact, positive control areas and, as such, *do* affect the VFR pilot.

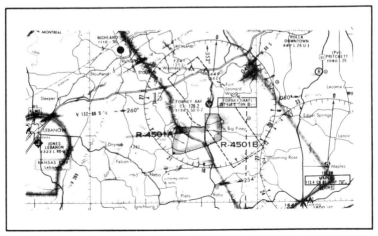

Fig. 8-5. A more or less rectangular transition area is outlined in soft pink wash on the Sectional charts.

Terminal Control Areas

Terminal control areas (TCAs) are those mushroom-shaped airspace monstrosities which the FAA has established in the vicinity of major terminal locations, ostensibly in order to control all aircraft within a given perimeter and, in turn, to improve the safety of all (Fig. 8-6). TCAs are the subject of much controversy, and for this reason we'll go into considerable detail to explain what they are all about.

In their simplest form, TCAs have been likened to upside-down wedding cakes because they consist of a series of tiers, increasing in size with altitude (Fig. 8-7). Originally all TCAs were designed to extend from the ground (in the center) to a maximum altitude of 7,000 feet agl. More recently the FAA has increased the tops of some to 12,500 feet, and plans are to extend the tops of all of them to that altitude.

The basic purpose of the TCA is to assure that all aircraft, from the mammoth 747 down to the lowliest Piper Cub be under ATC control while operating in the vicinity of a busy airport. Theoretically, they were supposed to improve safety and efficiency for all. In practice that is not exactly what has been happening.

Technically, a TCA is an area of positive control, meaning that *all* aircraft, no matter whether IFR or VFR, must maintain two-way

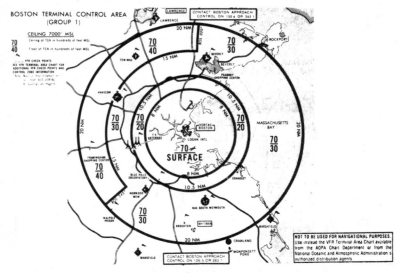

Fig. 8-6. A typical and fairly simple terminal control area.

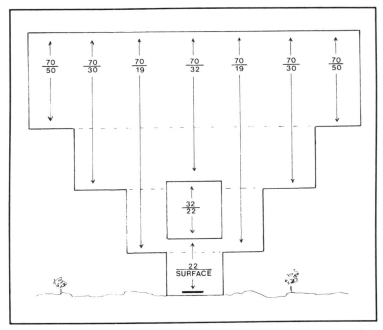

Fig. 8-7. A cross section of a typical terminal control area including a VFR corridor. All figures represent hundreds of feet msl (22 stands for 2,200 feet).

radio communication with the appropriate ATC facility, and must comply with instructions given by ATC controllers. This does not mean that a VFR aircraft should fly into a cloud, just because of a vector given by ATC. In a case like that the pilot is expected to inform ATC that he will not be able to comply with the instructions because he would not be in a position to remain VFR, and ATC is then expected to work out an alternate clearance based on information provided by the pilot. The TCA control rules do not require that all pilots be either IFR rated or that they operate IFR. The minimum license requirements for pilots are a private license, meaning that student pilots are not in a position to legally penetrate the outer limits of a TCA.

According to a new set of rules which was proposed more recently by the FAA, the minimum equipment requirement for operation within the TCAs includes not only two-way radio but also a transponder and an encoding altimeter. How strictly these rules, if enacted, will be enforced is a matter of local option. In one instance I found myself on my way to St. Louis when my transponder decided to give up the ghost. Upon contacting St.

Louis Approach I was informed that I would be permitted to land at Lambert Field, the controlling airport, but that I would not be permitted to take off again until the transponder was fixed. In that case I decided to opt for the Spirit of St. Louis Airport which is located outside TCA airspace.

The different levels and sections of TCAs are designated by horizontal boundaries and by vertical limits. Thus a certain section may include the airspace from the ground all the way up to either 7,000 or 12,500 feet agl, or it may have a base at, say, 3,000 feet, and go from that base up to the upper limit. All sections of TCAs (except for those few which underlie a so-called VFR corridor) extend upward to the overall upper limit of the TCA. But they vary as to base. It is thus possible to circumvent the inner portion of a TCA by staying below the outer sections' bases without having to communicate with ATC.

The trouble with most TCAs is that, in order to accommodate nearby airports and other local requirements, their airspace configuration is frequently extremely complicated, and some of them are so confusing that it requires considerable time of careful study to figure out where you can fly legally by staying out of the positive control airspace, and where it becomes necessary to start to communicate with ATC (Fig. 8-8). Most VFR pilots, finding one of these inverted wedding cakes along their route of flight, would be best advised to simply circumvent or to overfly the whole mess rather than try to figure out the specific restrictions applicable to each section.

An IFR pilot flying into or through a TCA can ignore the whole thing. As a matter of fact, as far as IFR pilots are concerned, TCAs do not exist (except that it is legally possible to fly IFR without a transponder—though not advisable—but it is not legal to fly into a TCA without one). It is the VFR pilot who must worry about them. He is expected to contact the controlling facility, usually Approach Control, on a frequency which is listed on the TCA charts at a point also indicated on those charts, a point which is usually some five or more miles outside the outer rim of the upper segment of the TCA. It is important here to state that VFR pilots who intend to operate into or through a TCA obtain one of the TCA charts for the area in question. Only these charts which are printed in a scale of 250,000 to one (Sectionals are printed in a scale of 500,000 to one and WACs are in a scale of 1,000,000 to one) contain all of the information needed to deal with a TCA when flying VFR.

A word of warning: While jets and other high-performance IFR aircraft are normally expected to enter and exit the TCA through

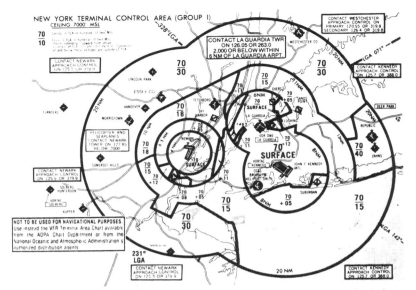

Fig. 8-8. The New York terminal control area is an example of how complicated they can get.

the top, this is not always the case. They quite frequently are found entering or exiting through the sides. Thus, simply flying around the perimeter of a TCA does not necessarily keep a VFR pilot out of danger. The other problem is that the airspace around the perimeter of the TCA and below its upper segments tends to compress a lot of VFR traffic into a limited area, often including huge numbers of student pilots. For this reason extreme vigilance is a necessity when operating anywhere in the vicinity of a TCA. Also, when flying VFR over the top of a TCA, always keep your radio tuned to the tower frequency to know in which direction airliners and other high-performance traffic are landing and taking off. This will help to anticipate where they would be likely to appear above the top of the TCA and give the VFR pilot an opportunity to stay well clear of them.

Positive Control Areas

The term *positive control area* means that *all* aircraft, regardless of whether operating IFR or VFR, must be in contact with and comply with ATC and its instructions. TCAs are one type of positive control area. Another type is all airspace above 18,000 feet msl. At these altitudes, the flight levels are designated not in thousands of feet but rather as Flight Level or, more generally FL.

Thus FL180 is 18,000 feet and FL 410 is 41,000 feet. Also, at all altitudes above FL180, the altimeter is always set to 29.92 rather than the current altimeter setting in the location over which the aircraft happens to be flying at the moment. The reason is that at that altitude obstruction clearance is no longer a problem (within the contiguous United States) and, since aircraft travel at very high speeds, the local altimeter setting would be meaningful for only a few minutes.

All aircraft flying at altitudes above FL180 must be flown by a pilot with at least a private license with an IFR rating, and they must be equipped with a transponder, an encoding altimeter, and a DME. Upon request by a pilot, ATC may make exceptions, but those are the rules.

This high-altitude positive control area includes all airspace from 18,000 feet msl up to and including 60,000 feet msl above all of the contiguous United States and most of Alaska, though in parts of Alaska its base is 24,000 feet. Above FL600 (60,000 feet msl) ATC has technically no jurisdiction over aircraft operations, but the only airplanes flying at such altitudes are military or an occasional Concorde SST. It is an area where the danger of a midair collision with some re-entering space garbage is greater than that of being confronted by another airplane.

The Continental Control Area

This is the airspace above 14,500 feet over all of the contiguous United States and over parts of Alaska, with the exception of any airspace within 2,500 feet of the terrain in mountainous areas. It is a plain and simple control area, meaning that IFR aircraft must be under continuous ATC control while VFR aircraft are not affected, except that they are expected to comply with the above-12,500-foot rule which calls for transponders and encoding altimeters. But since altitude recognition is impossible by radar unless the aircraft is equipped with an encoding altimeter, this is another one of those unenforceable rules which VFR pilots can and often do ignore.

A word of caution: A great amount of high-speed IFR traffic operate within the confines of the continental control area, VFR pilots should carefully adhere to the thousand-plus-five-hundred-foot cardinal-altitude rule to avoid the danger of a collision with IFR traffic which normally operates at even or odd thousand-foot levels, and whose pilots necessarily spend much time with the eyes concentrated on something inside the cockpit.

The continental control area extends from 14,500 feet up to 45,000 feet msl, though the portion above 18,000 feet is really meaningless, as here it coincides with the high-altitude positive control area.

Special-Use Areas

Special-use areas include those listed as *Warning, Alert, Training, Restricted* and *Prohibited* areas. Most are under one or another type of military control, and the rules and regulations governing them vary so widely that the important information with reference to each is printed on the aeronautical charts. Suffice it to say that all areas except those designated as either Restricted or Prohibited may be used by VFR and IFR pilots without obtaining any special permission. All that is needed is a considerable amount of extra vigilence.

In order to fly into or through restricted and prohibited areas during the hours during which they are in effect according to the information on (current) charts, the controlling agency will have to be contacted, or contact has to be made by using a flight service station as an intermediary, in order to obtain permission. When such permission is not available, unauthorized violation of such airspace can have rather drastic results, as some of them are used as training areas for heat-seeking missiles, something which the average light aircraft is ill-equipped to avoid.

There are also so-called *military operation areas* (MOA) in which the military may conduct IFR operations which are not under control of ATC. IFR traffic may be routed through such areas by ATC when safe separation can be assured. VFR traffic can use these areas at its own risk unless otherwise indicated on the charts.

The military also uses a variety of low-level training routes known by such names as *oil-burner routes, olive branch* routes and the like. They vary from time to time and are intermittantly published in the Airman's Information Manual. But since few pilots ever bother to read this overpriced publication, most of us usually don't know where they are and when they are in operation. Thus, if you spot some military aircraft operating at very low levels away from the vicinity of any military airport, the better part of valor is to make some kind of course correction and get away from there *in a hurry.*

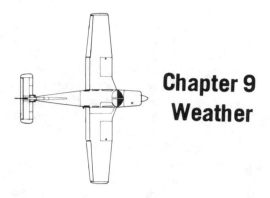

Chapter 9
Weather

Weather, like women, is a subject one can write about indefinitely without ever running out of variations. No single day's weather is ever like the weather on any other day or at any other location. But since weather is as important to aviation as the very ability to fly an airplane in the first place, we find it necessary to come up with some way to qualify it. Pilots who understand weather and learn how to deal with it and, especially, when and how to *respect* it, are likely to eventually end up dying in bed. Those who insist that weather is to be either fought or ignored are more likely to end up in the form of a statistic on some impersonal FAA report about the causes of aviation accidents.

What Weather Is

Weather consists of wide variety of ingredients which are the causes of differing results. Basically, *weather* is what happens when the air mass above the surface of the earth is exposed to variation in temperature. That's really all there is. High temperatures cause the density of the air to lessen, resulting in an expansion of the air mass being heated. Low temperatures cause the density of the air to increase, resulting in a compression of the air mass.

Logically, when a certain portion of the air is heated and thus made to expand, it increases in size and moves outward, producing wind. Conversely, the same happens when air contracts and the size of the air mass is reduced. Again, there is movement, meaning wind.

But this movement is affected not only by the expansion and compression process, but also by the rotation of the earth and its force of gravity. Without getting unnecessarily scientific, what happens is that the prevailing winds above the North American continent are usually westerly, while the rotation of the airflow around air masses is always predictable: Air flow around a high-pressure system, an area in which the air has been compressed to a degree of density greater than normal, is always clockwise. Conversely, the airflow around a low-pressure system, an area in which the density of the air is less than normal, is always counterclockwise. Within this circular movement of the air, the general tendency of the air relative to a high-pressure system is to flow away from the center of that system, while the air surrounding a low-pressure system tends to be sucked toward the center of that system.

All that is very interesting, but it doesn't mean a great deal to the average VFR pilot (and even relative to IFR flight it is not of overriding importance). What concerns us when we want to fly VFR from here to there is usually first of all, *where are the clouds and how much of them are there?* And then, *what are the winds at the altitude at which we want to fly?* And third, *are there any thunderstorms,* and if so, *is there a way to get around them?*

Clouds

Let's first look at the clouds. Clouds are moisture which has loosely coalesced in the air. At the altitudes at which most of us fly they are, more often than not, liquid moisture while, at higher altitudes, they tend to consist of ice crystals.

One way or the other, clouds represent a restriction to visibility. Inside a cloud the visibility is, for all practical purposes, *zero*, often restricted to the point that we can't even see the tips of the wings of the airplane we're flying. Since successful VFR flight depends on visibility, it is only logical that clouds must be avoided.

Now, most chapters on weather in aviation books and publications spend a lot of time and space explaining the difference between the various kinds of clouds, such as cirrus, cumulus, stratus and so on. That's of little practical use. (However, if you're interested, an excellent reference is *Pilot's Weather Guide,* TAB Book #2288). A cloud, no matter its name, is an *obstruction to visibility*, and there is no reason on earth why the pilot should care about its meteorological designation. As far as the pilot is concerned, there are *scattered clouds,* meaning clouds which represent a sky cover of between 10 and 60 percent. (Scattered clouds are *not* considered a ceiling.) As the term implies, there is lots of space between the individual clouds, sufficient to permit the pilot to climb up to an altitude above the clouds or to descend to a level below the bases of the clouds without much difficulty.

On a warm summer day the sky may be dotted with scattered (cumulus) clouds, indicating considerably turbulent conditions below the clouds, while the air above will be smooth and comfortably cool. These patchy clouds are the result of different parts of the surface below reflecting the heat of the sun to varying degrees. Underneath each cloud is likely to be a small rising current of air (a thermal), while between the clouds the air may either be smooth or may actually be moving downward in order to replace the air which is rising in the thermals; therefore the choppiness experienced when flying under such scattered clouds.

Broken clouds are clouds which cover anywhere from 60 to 90 percent of the sky. (Broken clouds are considered a ceiling, meaning that a condition of broken clouds with bases below 1,000 feet constitutes less than VFR minimums in controlled airspace.) When broken clouds are present and, assuming there is no additional cloud layer higher up, portions of the sky can be seen through the spaces between the clouds, and while climbing up the VFR conditions on top or descending from above the clouds to a level below their bases is still possible, it is somewhat more difficult than during scattered conditions.

VFR pilots, as a general rule, prefer to fly above layers of broken clouds where the air is smooth and cool, and where the visibility is usually better than down below. But staying above such clouds for any length of time requires that the pilot make repeated checks of the weather ahead. Broken clouds have a nasty habit of coalescing into a solid cloud deck, and while it is perfectly legal to fly VFR above a solid undercast, it can produce an embarrassing predicament if the pilot finds, once approaching his destination, that there is no way for him to come down. Therefore, before climbing to VFR conditions on top, the pilot should always make sure that there are ample breaks in the overcast at his destination, and that the forecast calls for improvement or, at the very least, no change, and above all, that there is ample fuel on board to permit flying a considerable additional distance in search of some kind of acceptable break in the undercast, just in case.

Overcast is the term used for clouds which cover 90 percent or more of the sky. Overcast can be thin, extending vertically for a hundred feet or less, with the sun actually visible through it, or it can be a solid mass of clouds extending upward for several miles. Under such conditions the VFR pilot has no choice but to stay below the clouds. There are occasions when breaks in a solid overcast are either reported or spotted by the pilot. Such breaks, even though they may reveal a patch of inviting blue sky, may easily turn into a sucker trap for the pilot who tries to use them in order to climb to VFR conditions on top. First of all, they are usually quite small, meaning that a climb through such a hole may call for a corkscrew-shaped flight path in order to keep from actually flying into clouds, and airplanes don't climb as well in a steep bank as they do when the wings are kept level. Second, unless there have been reliable and *recent* pilot reports with reference to the height of the tops of the clouds, the pilot may find that he breaks out at what he thought would be safely on top, only to find himself surrounded by huge shapes which, on all sides, extend upward for several additional thousands of feet.

Using such a break in the overcast in order to descend from VFR conditions on top to a level below the base of the overcast is somewhat less hairy. First, weather reports with reference to the base of the overcast are more easily available and can be expected to be reasonably reliable. Second, an airplane descends perfectly well, actually *better*, when the wings are in a steep bank. Thus a corkscrew-type descent is perfectly acceptable. The *bad* part of such an operation is that it is usually quite dark beneath the overcast and, coming from sundrenched conditions on top, we may find that it is difficult to clearly see in all that murk. A word of warning: When planning to use such a break in the overcast in order to descend through it, be sure that there is not another layer of clouds at a lower level. Quite often it is difficult to figure out whether those dark shapes which are visible through the break are actually the ground or some lower clouds. Look for anything that looks like a straight line. *There are no straight lines in clouds. Any* straight line is a sign of something man-made: a road, a railroad, the edge of a plowed field, or whatever. If there is a straight line anywhere, than what you are looking at is, in fact, the ground.

Thunderstorms

Thunderstorms are a menace. They, too, are given all kinds of fancy names by meteorologists, but from our point of view there are simply three types.

Isolated individual storms can be seen for miles and can easily be avoided by flying around the periphery on either side. But don't get too close. The turbulence associated with thunderstorms and occasional large-size hail have been known to occur as far as 20 miles from the actual center of the storm where heavy rain can usually be seen pouring out of the bottom of the clouds. It may cost a few extra minutes and an additional gallon or two, but it is well worth the added safety and comfort associated with a wider detour.

The second type is the so-called *squall line*. This is an entire family of thunderstorms, one next to the other, which usually occur in a line which often extends for several hundred miles. Such a line can represent a formidable barrier when it lies right across our intended route of flight. Any attempt to get over the top will usually prove fruitless, as the tops and even those occasional valleys between buildups tend to extend up to 30,000 or 40,000 feet, an altitude which is beyond the capability of most light aircraft, and one which should never even be considered without an ample supply of auxiliary oxygen on board.

Frequently, there are spaces between the individual cells which, at first glance, appear flyable. *Don't* try it unless you're a very experienced pilot (in which case, why are you reading this book?). Lightning can streak into those spaces or through them from cloud to cloud. The turbulence can easily be more than you're prepared to deal with. And there is always the better-than-even chance of encountering hail. If detouring around a squall line is impossible or impractical, the better part of valor is to land and wait. That wait may last until late into the night or the next day, but at least you'll be around in one piece to fly again. (For a more detailed discussion of playing tag with thunderstorms, see *Instrument Flying*, TAB book #2293.)

The third type is the so-called *imbedded thunderstorm*. This type is (or should be) of no consequence to the VFR pilot because the term imbedded means that it is hidden in a mass of clouds. Thus, in order to be faced by one of these imbedded thunderstorms, you would have had to be flying in clouds in the first place. Imbedded thunderstorms are a serious problem for IFR pilots, and when they are reported, pilots flying aircraft not equipped with airborne weather radar might do best to either drastically change their route of flight, or to abandon the flight altogether.

Wind

The second weather phenomenon of concern for the VFR (and the IFR) pilot is wind. Wind is virtually always with us, and at least

60 percent of the time—and that is a *conservative* estimate—the wind is unfavorable. Wind affects us in a variety of ways. At the surface, when we are ready to take off or land, the direction and velocity of the wind can complicate the job. Once airborne, it tends to try to blow us off our course and it has a serious effect on the ground speed, which, in turn, influences the amount of fuel needed in order to fly a given distance, and it either increases or reduces the range of the aircraft.

The statement that wind results more often in a negative effect than it does in a positive one may need a bit of explanation. Let's assume that we're flying for 300 miles into a 15-knot headwind in an airplane that flies at a true airspeed of 120 knots. With no wind, it would take that airplane two hours and 30 minutes to cover the 300 nautical miles. But the 15-knot headwind reduces the actual ground speed to 105 knots (120 minus 15). At that speed it will take the airplane two hours and 51 minutes and 25 seconds to fly the same distance. Now we turn around and fly the same distance the other way, this time taking advantage of a 15-knot tailwind. Now the ground speed is 135 knots and it will take the airplane only two hours and 13 minutes and 20 seconds to fly the distance. In other words, even though we flew an equal distance in both directions, and the wind component was the same each way, our exposure to the headwind was 38 minutes and five seconds longer than our exposure to the tailwind.

What this illustrates is that in view of the fact that we always end up with less ground speed when fighting headwind components, we simply end up spending more time with them than with tailwinds, even though over the long run the number of times when we're faced with either type of wind may be half and half. In addition, though we might think that a 90-degree crosswind would have no effect either way on the ground speed, it does, in fact, act as a headwind because, in order to stay on course, we have to correct our heading in the direction from which the wind is blowing. Thus, say we're flying along a 090-degree course with a 360-degree wind at 25 knots, making a true airspeed of 120 knots. We'll actually be flying a heading of 078 degrees and our ground speed will be only 117 knots. It actually requires at least five degrees on the tailwind side of a more-or-less right-angle crosswind to have zero effect on our airspeed. (The lower the airspeed, the greater the tailwind component has to be in order to produce a neutral result.)

The effect of wind on the relationship between course and heading is more subtle. Except when the wind is blowing from one side or the other at a quite considerable force, we're not really aware of the fact that the airplane is actually flying at an angle. As a matter of fact, most of the time we don't fly exactly in the direction in which the airplane is pointing. The airplane operates within an envelope of air. When that envelope of air is moving, it causes the airplane to move with it. Thus, if that envelope of air is blown to the left by the wind, the airplane inexorably moves to the left. If, in such an instance, we maintain a heading which coincides with the course, we'll end up a considerable distance to the left of where we have wanted to go.

To avoid confusion here, the *course* is the route we want to fly in order to get to our destination. The *heading* is the direction in which the nose of the airplane must point in order to fly that course. During flight training and as part of the written examination, there is a lot of talk about *true* course versus *magnetic* course and *true* heading versus *magnetic* heading. In either case, the relationship between true and magnetic is the number of degrees which must be deducted from or added to the true course or heading in order to arrive at the magnetic equivalent. True is referenced to *true north*. And magnetic is referenced to the *magnetic north*. All this may be very confusing to the student pilot working on his private ticket, but once we actually start flying for real, we always operate and think in terms of magnetic courses and headings, primarily because those are the ones indicated by the magnetic compass in the airplane, and they coincide with the compass roses shown around every VOR on our charts.

To give a few examples of the relationship between course and heading (true or magnetic, it makes little difference, as long as both are figured the same way. There is a slight difference because wind direction is always referenced to true north, but in practice it's not worth bothering with), here is what would happen under some admittedly extreme conditions:

Course: 180 degrees; TAS: 100 knots; wind direction: 240 degrees; wind velocity: 45 knots. *Result:* Heading to be flown: 203 degrees; ground speed: 70 knots.

Course: 060 degrees; TAS: 100 knots; wind direction: 350 degrees; wind velocity: 38 knots. *Result:* Heading to be flown: 039 degrees; ground speed: 80 knots.

Course: 170 degrees; TAS: 100 knots; wind direction: 350 degrees; wind velocity: 38 knots. *Result:* Heading to be flown: 170 degrees; ground speed: 138 knots.

Course: 090 degrees; TAS: 100 knots; wind direction: 315 degrees; wind velocity: 45 knots. *Result:* Heading to be flown: 071 degrees; ground speed: 127 knots.

Turbulence

In addition to its effect on ground speed and heading, wind is often responsible for turbulence. This is either the result of winds blowing over and around mountains and hills, or, usually at higher altitudes, of differences in wind direction and velocity at different levels. Except when reasonably close to the terrain in high mountain country, the turbulence resulting from wind is rarely severe enough to be of practical consequence. That doesn't mean that it can't be uncomfortable. It simply means that wind-generated turbulence is not likely to be of such severity to be structurally dangerous to the airplane or to make control of the airplane difficult. It's simply an unpleasant nuisance.

Precipitation

The third major ingredient of weather is *precipitation:* rain, snow, drizzle, freezing rain and all that good stuff. Rain, with the possible exception of unusually heavy downpours, is nothing much to worry about. It does reduce visibility to some extent, and, since the cabins of most light aircraft have an unpleasant habit of leaking somewhere, it can be annoying when water drips down your leg and into your shoes, but aside from that the airplane couldn't care less whether its raining or not.

Snow is a different story. When it's dry snow with the outside temperature comfortably below freezing, snow, too, is of no consequence, except that because of its high visibility, it gives an eerie feeling when we see those flakes coming at us at 100-plus miles an hour. But when the temperature is around freezing or a few degrees above, then snow tends to be wet and it will stick to the surfaces of the aircraft like glue, often turning into ice. This kind of snow can rapidly and drastically reduce the effectiveness of the lifting and control surfaces, raising the stalling speed in the bargain. It, and the worst of all types of precipitation, freezing rain, should be avoided and when they can't be avoided, the flight should be abandoned. Most light aircraft do not have a sufficient amount of extra power to draw on when the airfoils become deformed as the result of such precipitation, and there is always a good possibility that the stall speed will climb to a point at which the available power cannot produce sufficient airspeed to keep the airplane airborne. By that time it's too late. The time to look for a place to land is *before* things get marginal.

On the whole, weather is something that we learn to deal with as we gain more and more experience. But it must always be taken seriously. We should make it a hard and fast rule to always check on the weather before starting on a flight, and then to keep checking on it until the destination is clearly in sight. As long as we retain a reasonable amount of respect for the weather, we're not likely to have it ever get the better of us.

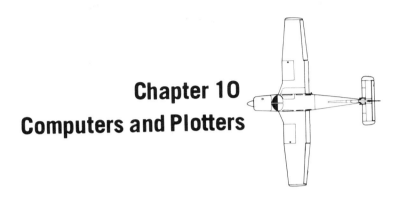

Chapter 10
Computers and Plotters

Manufacturers, anxious to develop products designed to separate the pilot from his money (and pilots, especially students and new pilots, are known to buy just about *anything* that sounds as if it might be useful), have devised a literally endless variety of gimmicks which are claimed to serve one or another useful purpose for the pilot. In fact, in addition to his charts and the instrumentation in the airplane, the pilot needs only two aids. One is known as a *plotter*, and the need for it tends to decrease with an increase in pilot-in-command time. The other is a *computer*, and in one form or another aviation computers continue to be used by all pilots throughout their flying careers.

The Plotter

As the name implies, a plotter is designed to be used when plotting a flight. For all practical purposes it is simply a ruler which, instead of being inscribed with inches or centimeters, shows statute and nautical miles relative to the Sectionals and WACs (Fig. 10-1). By placing the plotter on the chart along a straight line from point A to point B, it tells us the distance between the two points. That is the way in which it is used most of the time.

But in addition to being a ruler, the average plotter has a sort of half-moon-shaped extension, the usefulness of which is primarily limited to helping find the answers to some of the navigation

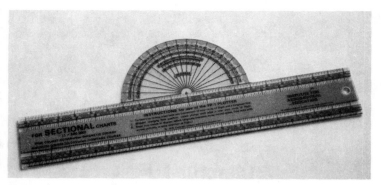

Fig. 10-1. A typical plotter.

questions, in the written examination. On this semi-circular portion lines are inscribed emanating from the center of the circle, each one identified by a number designating degrees. By placing that center of the circle (usually a hole through which a pencil point or needle can be pushed) over a longitude line (the north-south line) on the chart and by then rotating it until the edge of the ruler portion coincides with or parallels the route of flight, the number on top of the longitude line will indicate the true course (not magnetic course). But since we don't much care about the true course, and since virtually any flight will cross or pass near at least one VOR, and since the VOR radials shown on the charts are magnetic directions, it is usually a lot simpler and more useful to get the course direction from one of the VOR compass roses. Personally I don't remember having used a plotter for any purpose other than to determine the distance between two points since the day I passed my private written, and that was a *long* time ago.

The E6b Computer

Don't ask me why it's called E6b. That designation is some sort of hangover from the military and has absolutely no meaning whatsoever. It simply stuck. The E6b computer is, in fact, a circular slide rule (Fig. 10-2). By rotating the disk which is the upper portion of the computer, the data imprinted on it can be placed opposite other data, and then, by comparing one set of data to the other, we can determine all kinds of facts which we may need to know.

Pilots most frequently use the E6b to determine true airspeed, to find out how long it will take to fly a certain distance at a given ground speed, and to find out how much fuel it will take to fly that distance. The computer is also capable of telling us the density altitude under a variation of conditions, the time available for flight in hours and minutes with a given amount of fuel at a certain fuel flow, and, some of the more expensive versions include a wind-correction-angle computation capability on the back. Let's look at one thing at a time.

The routine is really quite simple. To determine how long it will take to cover a certain distance at a given ground speed, we put the black indicator marked 60/GPH/MPH on the rotating portion of the computer opposite the figure representing the ground speed (Fig. 10-3). The ground-speed figure can be found on the stationary portion of the computer in two-digit figures to which a zero must be added. Thus, if our expected ground speed is 130 knots (or miles), we put the indicator opposite 13. Now we go around the outer perimeter of the stationary portion of the computer to find the distance to be flown. Let's say the distance is 240 nautical (or statute) miles. Now by comparing the time figures printed on the inner row of the moving portion of the computer, we find that the figure 1:50 is just about opposite the 24 on the other rim. This means that at 130 knots it will take a tad over one hour and 50 minutes to cover 240 nautical miles. Now, if we want to know how much fuel that will take, we move the indicator (60/GPH/MPH) opposite the number on the outer rim which coincides with the amount of fuel being burned per hour (Fig. 10-4). In this case, assuming the fuel flow is nine gallons per hour, we place the indicator opposite 90 which, in this case, represents 9. Then we check where the time figure, 1:50, on the inside row of the circular

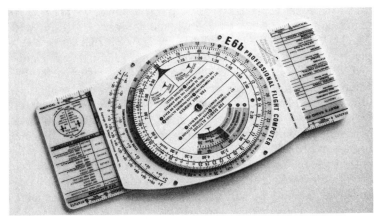

Fig. 10-2. The front side of a typical E6b computer.

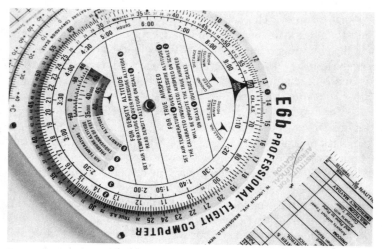

Fig. 10-3. In figuring time/distance problems, the indicator is placed opposite the ground speed and the time can then be found opposite the distance.

portion of the computer points in relation to the figures on the outer rim. We find that it points halfway between 16 and 17, meaning that it will take 16.5 gallons to fly that distance in that time. All time, distance and fuel problems can be solved with a reasonable amount of accuracy in this fashion, and all it takes is maybe a half an hour of practice, after which using the computer becomes second nature.

To determine true airspeed, things get just a trifle more complicated. First we set the outside air temperature, marked AIR TEMPERATURE °C on the moving part of the computer, opposite the altitude as displayed on the aircraft altimeter. This altitude figure is in the next inner row and is marked PRESSURE ALTITUDE THOUSANDS OF FEET. Let's say that the temperature on our OAT indicator is shown in degrees F., and that it shows +25. But we need degrees C. So we look at the bottom of the E6b where we find a scale comparing Fahrenheit and Centigrade. This tells us that +25°F. equals approximately −4°C. Now, assuming we're flying at 15,500 feet, we place the line representing −4 opposite the space halfway between 15000 and 20000 which represents 15,500 feet. We can now read the true airspeed on the outer scale of the stationary portion of the computer where it will be opposite the figure representing the indicated airspeed which is shown on the outer scale of the movable portion of the computer (Fig. 10-5). Thus, in this example, if the indicated airspeed is 155 knots, the TAS will hover just around 200 knots.

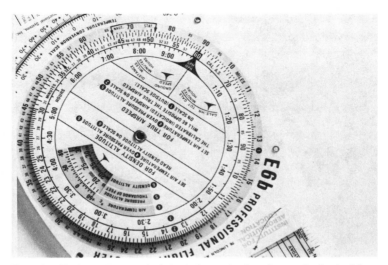

Fig. 10-4. To figure fuel needed, the indicator is placed opposite the fuel flow and the fuel needed can then be read opposite the time en route.

A similar procedure is involved in determining the density altitude. In this case we again place the air temperature over the altitude or airport elevation, and the indicator at the bottom of that window will show the approximate density altitude (Fig. 10-6).

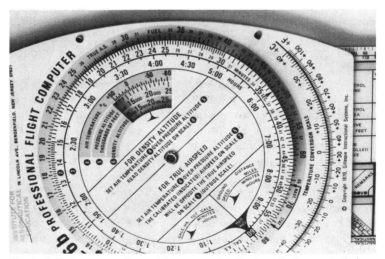

Fig. 10-5. To figure true airspeed the temperature is placed in the small window opposite the altitude and the true airspeed can then be read opposite the indicated airspeed.

151

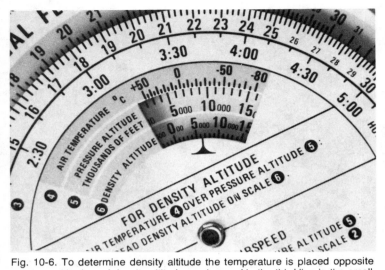

Fig. 10-6. To determine density altitude the temperature is placed opposite pressure altitude and density altitude can be read in the third line in the small window.

Assuming, for example, that the airport elevation is 5,000 feet and the temperature is +20°C., we will find that the density altitude is roughly half way between 6,000 and 7,000 feet, or somewhere near 6,500 feet.

Now let's turn the thing over. The back portion of the computer consists of another rotating circular position, the stationary portion of the computer itself, and a slide which can be moved up and down and the indications on it can be seen through the transparent portion of the circular moving part (Fig. 10-7). All of this is for the purpose of determining wind-correction angles and the ground speed which can be expected under given wind conditions. In practice few pilots ever bother to use this part of the computer because it somehow seems to be more trouble than it's worth. But it does come in handy when taking written examinations. Here is how it works:

The curved lines on the slide which can be seen through the transparent window represent two knots of wind velocity per line. We start off by placing the circular portion in such a way that the wind direction is under the TRUE INDEX mark. We now take a pencil and mark the velocity up from the center grommet with a dot (Fig. 10-8). We now rotate the movable circular portion until our true course is under the TRUE INDEX mark. We then slide the movable slide up (or down) until the pencil dot is on the line which

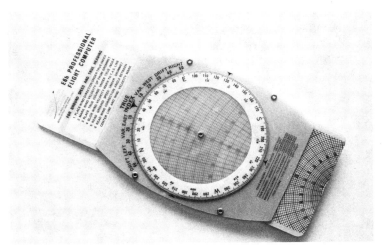

Fig. 10-7. The backside of a typical E6b computer, used to figure wind-correction problems. To figure the wind correction angle and ground speed, first place the wind direction under the TRUE INDEX mark.

coincides with the TAS (Fig. 10-9). The position of the center grommet will now show the ground speed and the numbers along the curved lines on the slide will indicate the correction angle. If it is to the left of the center line, the number has to be deducted from the true-course figure. If it is to the right of the center line, it has to

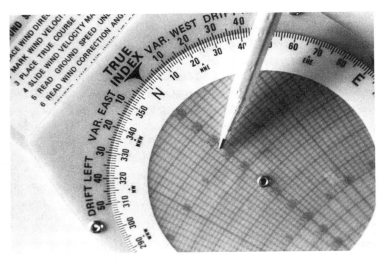

Fig. 10-8. Now, using a pencil, mark the wind velocity upward from the center grommet.

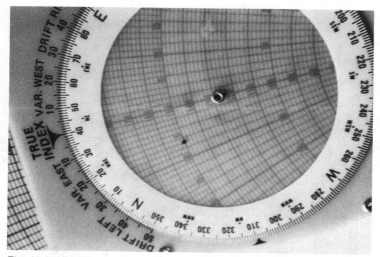

Fig. 10-9. Now move the course under the TRUE INDEX mark and movable slide until the pencil dot is on the line representing true airspeed. The position of the grommet now represents the ground speed and the position of the dot shows the direction and degree of the needed wind correction angle, in other words, the heading.

be added to that figure. Here is an example: The wind is out of the north, 360 degrees. The true course is 040 degrees. The wind velocity is 28 knots. The TAS is 150 knots. The position of the grommet now shows that the ground speed will be 128 knots and that the wind correction angle is seven degrees to the left, meaning that we should be flying a heading of 040 minus 7 or 033 degrees. This will be the *true* (not magnetic) heading. To come up with the magnetic heading we still have to deduct easterly variations or add westerly variation. This last step can, of course, be eliminated by using the magnetic instead of the true course in the first place. The result will include a small error, but that is not likely to be sufficiently significant to be of consequence. The reason for the error is that winds are always reported in directions relative to *true* north.

E6b computers come with and without the wind-correction slide. Some are plastic, some are metal, some are cheap and some quite expensive (Fig. 10-10). For the average general-aviation pilot and his day-to-day use of the computer, the small, cheap versions are just as good as the expensive ones.

As can be seen by the above examples, these mechanical slide rule-type computers produce results which are approximations. It is impossible to get really precise answers with them. But precision in flying is becoming increasingly important with the high price of today's aviation fuel. For this reason the new types of electronic computer-calculators designed specifically for aviation will, in the long run, return their somewhat higher cost in terms of fuel savings.

Electronic Computers

The most popular among these is being marketed at the time at which this book is being written by two companies, Jeppesen (under the name of *Avstar*) and Cessna (under the name of *Sky/Comp*). Both are identical in function as well as price ($69.95 in 1980). This computer, which is small enough to fit into the average shirt pocket, does everything that the E6b is capable of, but with much greater precision, plus being able to perform a number of other functions not available with the mechanical slide rule-type computer (Fig. 10-11). Another, and more expensive type ($119.50 to $299.50 depending on certain optional functions) is the Navtronic 1701 (Fig. 10-12) which comes in four versons, the 1701, the 1701r, the 1701t, and the 1701tr. Of the two the latter is somewhat easier to use regardless of the specific model, but both are excellent pieces of equipment.

The principle of operation employed in these electronic aviation computers is this: the computer has been preprogrammed to perform certain computations automatically. All the pilot has to

Fig. 10-10. Aviation computers come in many varieties. For day-to-day use the simple ones are quite acceptable, though no wind problems can be solved with them.

Fig. 10-11. The Jeppesen Avstar and the Cessna Sky/Comp are the electronic equivalents of the E6b, though they are capable of solving a considerable number of additional aviation-related problems.

do is to tell the computer certain basic established data and then what type of program the computer is supposed to use. This input of information and commands is accomplished through the use of a variety of keys, 40 on the Avstar (or Sky/Comp) and 30 on the Navtronic. Each type of computer comes with a complete and detailed instruction booklet; it would therefore be a waste of space here to go into complete detail of what must be done in order to obtain the answers to all possible questions.

But in order to present a reasonably clear picture of the degree of precision which is available with electronic pocket computers, let's just use the same problems which we described above with reference to the E6b.

Problem: How long will it take to fly 240 nm at a ground speed of 130 knots? Routine using the Avstar: Press the *T-S-D* key to tell the computer that the time-speed-distance program is to be used; enter 240 and press *Dst/Fuel* to identify that figure as distance; then enter 130 and press *GS/FPH* to identify that figure as ground speed. Then press *Comp* to tell the computer that we are ready to start computing, and then press *Time*, to let it know that the answer we want is time. The result: The display window will show

1-50-46, which means that it will take us one hour, 50 minutes and 46 seconds to fly that distance at that speed. (The answer with the E6b was 1:50-plus.)

Routine using the Navtronic: Press *TIME* to tell the computer that we're looking for a time-type answer. A little red light now lights up to command the next entry. It calls for *DIST*. We enter 240. The red light now calls for *SPD* and we enter 130. The result: The display window will show 01 50 46. (Apparently the two agree with one another.)

Problem: How much fuel is used in the above flight, assuming fuel flow at nine gph? Routine using the Avstar: Press *T-S-D*

Fig. 10-12. The Navtronic 1701 is a more sophisticated and in some ways easier to use aviation computer.

because technically this, too, is a time, speed, distance problem. Press 1 then *T:* then 50 then *T:* then 46 which will show up in the display window as 1-50-46, the time en route. Now press *Time* to assure the computer that this figure refers to time. Now press 9 and then *GS/FPH* to tell the computer that we're talking about 9 gph. Then press *Comp* and *Dst/Fuel*. The result: 16.6, meaning that the fuel used will be 16.6 gallons. (The E6b came up with 16.5.).

Routine using the Navtronic: Press *FUEL* to tell the computer that that is what we're after. The red light tells us that the computer wants to know *RATE*. We enter 9. Now the red light asks for *TIME*, and we press 1 then ·/: twice then 50 then ·/: once then 46 and *ENT* to enter that into the computer. The result: 16.6.

Problem: Determine TAS when flying at 15,500 feet with an OAT of −4 degrees and an IAS of 155 knots. The E6b came up with something around 200 knots.

Routine, using the Avstar: Press *Alt/AS*, then enter altitude by pressing 15500 and *P Alt*, then the outside air temperature by pressing 4 then +/− then *I°C*, and then the indicated airspeed by pressing 155 and *CAS*. To get the result we press *Comp* and *TAS*. The result: 198.

Routine using the Navtronic: Press *TAS* to identify the desired result. The red light asks for *PRES ALT* and we enter 15500. The red light now asks for *AIR TEMP °C* and we enter it by pressing 4 and +/−. The red light now asks for the indicated (or calibrated) airspeed by appearing next to *CAL AIR SPD*. We press 155. The result: 201. Apparently the two do not completely agree with one another. Frankly, I don't know which one is right and which one is not, but, being within three knots of one another, the difference is not sufficient to be of importance.

Problem: Determine the density altitude at an airport with a 5,000-foot elevation when the temperature is +20°C.

Routine, using the Avstar: Press *Alt/AS*, then enter the airport elevation by pressing 5000 and *P Alt*. Then enter the temperature by pressing 20 and *T°C*. Then press *Comp* and *D alt*. The result: 6712. (With the E6b the result was shown as around 6,500 feet.)

Routine, using the Navtronic: Press *D ALT* to identify the desired data. The red light asks for *PRES ALT* and we enter the elevation, 5000. The red light now asks for *AIR TEMP °C* and we enter 20. The result: 6712. (Again the two computers don't agree completely, but the difference is inconsequential.)

Problem: The wind is blowing from 360 degrees at 28 knots. Our course is 040 degrees, and our TAS is 150 knots. What is the heading to be flown and the ground speed which will be made good?

Routine using the Avstar: Press *Wind*, then enter wind direction by pressing 360 and *W Dir*. Then enter wind velocity by pressing 28 and *W Spd*, then enter course by pressing 40 and *Crs*, followed by the true airspeed, 150 and *TAS*. Now press *Comp* and *Hdg*. Result: 33 or 033 degrees. Then press *Comp* again followed by *GS/FPH*. The result: 127. In other words, the ground speed will be 127 knots on a heading of 033 degrees. (The E6b came up with 128 knots and a heading of 033 degrees.)

Routine using the Navtronic: Press *GS+HD* to activate the appropriate program. The red light asks for *WIND DIR* and we enter 360. It now asks for *COURSE* and we enter 40. It now asks for *WIND VEL* and we enter 28. The next request is for *TRU AIR SPD* and we enter 150. The result: 127 033. This time both computers are in complete agreement.

We could go on examining all the varied capabilities of both types of computers, but that could practically fill an entire book. Suffice it to say that both can also be used as calculators in balancing out your checkbook or deciding whether the "large economy size" is, in fact, a better deal than several small ones.

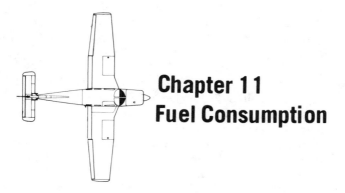

Chapter 11
Fuel Consumption

In the old days, when aviation gasoline was cheap and plentiful, few pilots paid much attention to the questions involved with conserving fuel. Today this has changed. At a fuel cost of over $2 per gallon, we simply can no longer afford to blithely bore holes in the sky, operating our aircraft at some power setting approximating 75 percent of power. Today it is important that every pilot understands clearly which factors govern the amount of fuel being burned in order to cover a certain distance in the available amount of time. There are three types of controls available to the pilot flying aircraft equipped with constant-speed propellers, and two types to those flying aircraft with fixed-pitch props. These are the *mixture*, the *manifold pressure* (throttle) and the *rpm controls*.

Mixture

Cruising with the right mixture, regardless of altitude, reduces fuel consumption, is kinder to the engine, increases engine life and time between overhauls, and prevents spark plug fouling. The mixture is the blend of air and vaporized gasoline that is burned in the engine, causing the pistons to move and thus the prop to turn. The perfect proportion is 15.2 pounds of air to each one pound of fuel. At this proportion both air and gasoline are completely consumed with only so-called pure combustion products, carbon dioxide and water, remaining. Mixtures as rich as nine to one and as lean as 20 to one are capable of combustion, but they leave quantities of other leftovers (products of disassociation) of which carbon monoxide is the best known.

Rich mixtures contain an excessive amount of fuel, while lean mixtures are composed of too much air. Though both are theoretically non-ideal, they may be selected to produce particular results. Engine cooling as well as achieving maximum power call for a richer-than-ideal mixture and in the day-to-day operation of light aircraft rich mixtures are used quite frequently while lean ones are thought by many to be detrimental to the health of the engine, which is, in fact, not necessarily the case.

The only way to accurately determine the composition of the mixture is to read the temperature of the exhaust gases. These gases are hottest when the mixture is ideal, and cooler the farther away from ideal the mixture gets in both the rich and lean directions. In order to permit the pilot to monitor the exhaust-gas temperatures, the aircraft must be equipped with an exhaust-gas-temperature gauge (EGT). In the absence of such a gauge all the pilot can do is lean until the engine starts to run rough and then to enrich just enough to restore smooth operation.

Reciprocating engines, whether fuel injected or carburetor equipped, contain a so-called automatic enrichment valve that enriches the mixture at idle and at full throttle, leaning it somewhat during the balance of the operating range. At idle this neutralizes the bad effects on combustion of unscavenged exhaust gases, and at full throttle it helps to cool the engine, protecting it from possible damage due to detonation.

This automatic system is adequate only at sea level under standard atmospheric conditions. It does not react to changes in air density or temperature. This must be accomplished by the pilot through the use of the mixture-control lever.

The full-rich mixture setting is designed to give adequate cooling during takeoff and climb. At full power it is therefore often necessary to maintain an exhaust-gas temperature some 200 or more degrees F. on the rich side of peak. This is accomplished by that enrichment valve. But since it operates only when the throttle is shoved all the way in (or at idle), pilots who think that they are babying their engines by reducing power slightly during the climb are actually accomplishing the opposite.

Reciprocating engines are designed to operate at this full-rich setting when producing 75 or more percent of power, while at any setting below 75 percent, be it during climb, cruise or descent, the mixture may be leaned (within the limits stated by the manufacturer in the operating handbook). What is considered conservative by many, and what is often taught by flight instructors who frequently don't seem to understand what mixture control is all about, namely not to start leaning at any altitude below 5,000 feet, is not conservative at all. It increases fuel consumption, reduces the range of the aircraft, tends to increase spark plug fouling and shortens the TBO.

Leaning without an EGT gauge is, at best, an inexact science. The best one can do is to lean slowly until the engine runs rough (or, in aircraft equipped with fixed-pitch props, until there is a noticeable drop in rmp)and then to enrich again slowly until smooth operation is restored. (In fixed-pitch-prop aircraft it is not advisable to continue to enrich until maximum rpm is again reached. Maximum rpm does represent best power, but in practice produces only about a two-percent increase in speed at a 14 to 15-percent fuel-consumption penalty.) With today's cost of fuel, the relatively modest investment involved in installing an EGT gauge will soon be returned in the form of meaningful fuel savings.

For greatest economy it is advisable to cruise at the leanest mixture permitted by the engine manufacturer. Depending on the individual engine, this may be 25 degrees on the rich side of peak, at peak, or even 25 degrees on the lean side of peak. It is important to be aware of the manufacturer's recommendations for the particular engine being flown, as he has determined the safe limits which should not be exceeded by the pilot. Once the desired mixture has been obtained, it remains correct *only* as long as the combination of rpm, manifold pressure, density, altitude and temperature continues unchanged. As soon as there is any variation, the mixture changes too, though these changes may often be too small to warrant adjustment.

If atmospheric conditions along the route of flight require the use of carburetor heat, the mixture must be readjusted as carburetor heat causes enrichment which in no way improves or affects the ice-melting capability.

The degree to which fuel economy is affected by mixture is quite dramatic. To take an extreme example: If an aircraft were flown (stupidly) at full rich at 7,500 feet, it would consume twice the fuel of one flown at peak EGT, doubling the fuel bill and cutting the aircraft's range by half.

Manifold Pressure and RPM

Aircraft equipped with variable-pitch propellers offer additional fuel-saving options in the form of the rpm and manifold-pressure controls. The function of these two controls has been explained in detail in another chapter. Suffice it to say here that rpm represents the speed with which the engine turns, in other words, the number of times the pistons perform their complete four-stroke cycle within a given period of time. Obviously, the faster the engine turns, the more frequently fuel is injected into it and, in turn, more fuel is being consumed. Logically, therefore, fuel can be saved by reducing the rpm to an acceptable minimum and the manifold pressure to the acceptable maximum to arrive at a power setting that produces the desired percent of power. As an example, some engines can safely be operated at a miserly 22 *in hg* of manifold pressure and 1,900 rpm.

Caution must be exercised when switching from climb to cruise power or from cruise to descent. After a climb at, say, 2,500 rpm and 25 *in hg* of manifold pressure, when it comes time to change to a cruise setting of 1,900 rpm and 22 *in hg* of manifold pressure, never reduce the rpm first. This would result momentarily in a 1,900/25 setting that could prove ruinous to the engine. The simple basic, but important, rule is: *When reducing power, first manifold pressure, then rpm. When increasing power, first rpm, then manifold pressure.* (As a memory jogger, think of it this way: Correlate the M in manifold with *minus* and the P in rpm with *plus*. Reducing power calls for minus, or manifold pressure first. Increasing power calls for plus or rpm first.) This is also the real reason for changing to full rpm before landing. Just imagine an 1,800-rpm setting during descent, followed by the decision to go around. Shoving in full throttle could produce 1,800 rpm and 29 *in hg* of manifold pressure, which very likely would produce catastrophic engine failure within *less than a minute*.

Speed and Altitude

A simple change in the way we look at speed can save huge amounts of fuel. Don't think how fast the airplane can get you to your destination, but rather how much time you have available to get there. The trick is to always select the *slowest* speed within the time available.

Three basic speeds which vary with aircraft and altitude are of importance here. One is the *maximum cruise speed*, for most aircraft 75 percent of power up to the altitude at which 75 percent coincides with full throttle. This is the speed which gets us there fastest, using a high degree of fuel flow and resulting in a reduction in range. The second is the *best-range speed*. This is the one that gets us the farthest on the least amount of fuel. For most aircraft this is approximately 40 percent above the best-angle-of-climb speed and it should not be confused with the third, the *best-endurance speed*. This latter permits us to stay aloft longest, regardless of the distance flown. It is somewhat below the best-rate-of-climb speed. In practice, this is the speed to be selected when ATC announces a delay or when we arrive VFR on top of a solid undercast at our destination and the forecast calls for clearing within the hour.

For cruise, the best-range speed represents the quintessence of economy. It is rarely given in manufacturer's performance charts because it falls somewhere around 45 percent and the resultant speed is too low to look good to the buying public. Since true airspeed increases with altitude in relation to the indicated airspeed, the ideal altitude to achieve absolute optimum range is that at which full throttle produces 45 percent of power. This is usually too high to be acceptable without auxiliary oxygen, and flying at a more comfortable 10,000 or 12,000 feet will result in only a slight increase in fuel consumption. It must be remembered that the increase in fuel flow at higher power settings is quite out of proportion to the percentage of speed gained. For instance, 55 percent of power will result in 10 to 15 percent less speed than 75 percent of power, but the fuel saving will be between 25 and 30 percent.

Of course, in figuring the advantages of flying at higher altitudes, the effect of wind must be taken into consideration. For maximum economy, both head- and tailwinds call for adjustments in the cruising speed. With a tailwind it is more economical to reduce power and maintain the ground speed that would have been achieved without wind, thus exposing the aircraft for a longer time to the beneficial effects of the tailwind. Conversely, it is advisable to increase speed when flying into a headwind, reducing the time of exposure to its detrimental effect. As a rule of thumb, these decreases and increases should be equal to about one fourth of the head- or tailwind components.

When the winds are in excess of 15 to 20 percent of the aircraft's true airspeed, it is usually better to change altitude to avoid headwinds or to take advantage of tailwinds. At those velocities the advantages to be gained by flying at high altitudes would be cancelled out by the headwind, while the economical disadvantages of flying low would become meaningless in view of the added push provided by the tailwind.

Pilots who pay close attention to the subjects discussed above will find that their annual fuel consumption may be reduced by as much as 25 percent while flying as much as ever with only a minimal increase in the time spent aloft, covering the same distances as before.

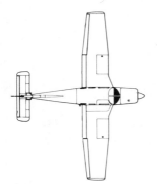

Chapter 12
Avionics for Communication and Navigation

Avionics, a contraction of the words aviation and electronics, are not needed to permit an airplane to fly. On the other hand, there is very little airspace left in the United States today where an airplane without at least a minimum of avionics can operate legally, not to mention safely and efficiently. Avionics, often referred to as *black boxes*, are our means of communicating with persons on the ground (or in other aircraft) and of navigating without reference to visual landmarks on the ground. Today it would be difficult indeed to find an airplane, *any* kind of airplane, that does not have at least some avionics equipment on board (Fig. 12-1). The type of avionics equipment available today is so incredibly varied in terms of function, sophistication and, of course, cost, that we could fill several books of this size, examining each in complete detail. For this reason we'll limit ourselves in this chapter to those black boxes which are the most useful in relation to single-engine aircraft and light twins.

Communication

Communication radios, commonly referred to as *com* systems, are simply two-way radios, much like the CB (Citizen's Band) radios which were recently popular on our highways. They consist of a receiver and a transmitter, both usually combined in the same box and using the same antenna (Fig. 12-2). They are known as *simplex* systems, meaning that it is possible to either

Fig. 12-1. A relatively full complement of avionics is needed in order to effectively use today's airspace (courtesy of Piper).

transmit *or* to receive, but impossible to perform *both* funtions at the same time (such as is the case with the normal telephone which is, in fact, a *duplex* system). When turned on, the system will receive all transmissions which are transmitted over the particular frequency to which the system is tuned. In order to transmit, a press-to-talk button on the microphone has to be depressed. It cuts out the reception capability and causes the system to act as a transmitter. Because these systems combine the receiving and transmitting capabilities, they are referred to as *transceivers* (a contraction of transmitting and receiving). Today all aircraft transceivers used within the United States are operating on the VHF (very high frequency) wave band, the communications frequencies being restricted to the range from 118.0 MHz (megaHertz or megacycles) to 135.975 MHz, with the minimum spacing between frequencies in use being .025 MHz. This means that within that range there are 720 different frequencies on which it is possible to communicate.

Most com systems are designed to be tuned to one frequency at a time such as, say, a tower frequency, and then to use that frequency for transmission as well as reception. Some of the older models, such as the venerable Narco Superhomer, few of which are still in use, used to have a coffee-grinder-type handle to tune in reception frequencies, while the transmitting frequencies were limited to a few for which a special crystal had been installed. In such a system, the pilot would tune to 122.5 for transmission to a tower, while his receiver might be tuned to 118.7 or whatever frequency was published for the tower. All this was quite awkward, and it's a good thing that the state of the art has advanced to a point where it is no longer needed.

With the solitude of flight being one of the more delightful aspects of flying, why do we want to talk to those guys on the ground in the first place?

Part of the answer is that without talking to anyone, our aerial activity would be limited to flying from one uncontrolled airport to another in weather which we know to be amply VFR all the way. Taking off from or landing at a controlled airport (any airport with an operating control tower) involves talking to the tower controllers, making on-board communication equipment *mandatory*.

The other part of the answer is that there is no way for us to keep track of our en-route weather on flights of any duration, except by intermittantly listening to weather reports or calling ground stations to ask for the weather at our destination or at point along the route.

Futhermore, flying IFR is impossible (or, at least, illegal) without the appropriate communications capability, and even VFR pilots will find it advantageous, even at uncontrolled airports, to be able to talk to unicom in order to obtain information about the runway in use, known traffic in the area, and so on.

Many of the newer transceivers are designed to be able to store one or several communications frequencies in their memory in addition to the one which is currently being used. The reason is simple. With today's relatively complicated airspace control, even a VFR pilot approaching a major airport may be called upon to use a series of frequencies in quick succession. First he'll be talking to approach control, then he'll be turned over to the tower, and once clear of the active runway, he is expected to tell ground control where he wants to go on the airport. All this switching of frequencies can get to be a major pain in the neck, especially if the weather is bumpy and if there is a lot of traffic to look out for.

For pilots flying IFR the situation is likely to be even worse. On even a relatively short flight a pilot might be called upon to use a dozen or more different frequencies, and always having the next one ready tends to reduce pilot workload. Furthermore, when asked to switch to a new frequency, it is always a comforting idea to keep the last one in reserve. It happens not too infrequently that we find that we can't make contact on the new frequency, but as happens equally often, by then we have forgotten what that last

Fig. 12-2. A VHF transceiver (courtesy of Narco).

frequency was. But if we stored it, all we have to do is switch back and tell that last facility our problem.

Because communication capability is such an unavoidable necessity these days, most halfway decently equipped aircraft will have dual transceivers on board. This not only provides the ability to keep at least one spare frequency handy for use when needed, it also means that if one of the radios should decide to develop some sort of trouble, we still have the other one to get us to where we want to go.

Aviation communication radios cannot legally be used in the casual manner in which all kinds of nuts are using those CB transceivers. With the single exception of the unicom frequencies (which can legally be used to order a taxi or to ask someone on the ground to call the pilot's wife or husband in order to be met at the airport upon landing), all aviation-frequency use is limited to strictly aviation business. And because a lot of people, be they controllers or pilots, will often need to have access to a limited number of frequencies without unnecessary delay, all users are urged to keep their transmissions as short as possible. For this reason, some generally accepted aviation phraseology has been developed over the years, phraseology which is designed to min-imize misunderstandings and to keep the number of words being spoken to a minimum. There is no law that says that you *can't* use just plain English, but failing to make use of that accepted spoken "shorthand" automatically brands the speaker as a novice.

To give just a very simple example: the pilot wants to tell ground control that he is ready to taxi to the runway for takeoff. Using a normal everyday speech pattern he might say: "Hello, Ground Control. This is November One Two Three Four Alpha. I'm here at XYZ Aviation and I wonder if you could clear me to taxi to the active runway, over." The same message could be accomplished by: "Ground, Three Four Alpha, XYZ, taxi to takeoff. Over." It takes less than half the time and leaves less margin for error.

Another advantage to be derived from the way aviation communications are handled is that every experienced pilot and controller will always place each piece of information to be transmitted into a certain, firmly established, sequence. And certain words or phases can be used only under certain conditions. Thus, no controller is *ever* permitted to use the word "takeoff" at any time other than in conjunction with the words "cleared for . . . " Thus, even if there is a lot of static or the transmission is garbled for one reason or another, if the word "takeoff" is heard, it means that the aircraft is cleared for takeoff. A tragic result of the misuse of that word was the collision of two 747s on the airport at Tenerife some years ago. The tower, or possibly ground control, told one 747 to taxi to the "takeoff position" or "takeoff end" of the runway, while the other was using the active runway to also taxi in that direction. Apparently the transmission was less than clear. In any case, the pilot, having heard only the word "takeoff" clearly, decided that he was cleared for takeoff. With the visibility at the time down to not much over 1,000 feet RVR (runway visual range), he started his takeoff run and by the time he finally saw the other 747 facing him on the active runway, it was too late for evasive action. The collision killed hundreds of people.

The same is true with reference to the consecutive order in which information is transmitted. More often than not we tend to miss certain portions of a transmission, but knowing that what we did hear was the important part, we can safely ignore the rest for the time being. It takes a bit of practice, but we soon learn to listen for what we need to know, and to ignore whatever is not of immediate importance.

In addition certain words and phrases have become accepted usage, either because they reduce a longer sequence of words to a single phrase, or because they reduce the chance for misunderstanding:

Acknowledge: Please confirm that you have received and understood the message.

Affirmative: Yes.

Go ahead: Proceed with your message.

How do you read?: Can you hear and understand me all right?

Ident: Press the ident button on your transponder.

I say again: I repeat what I have already said.

Negative: No. Incorrect. Or, permission not granted.

Out: My message is finished and I am expecting no reply.

Over: My message is finished but I am expecting a reply.

Read back: Please repeat the message in its entirety.

Roger: I have received and understood your last transmission.

Say again: Please repeat what you said.

Squawk (followed by a number): Turn your transponder to a new frequency as per the number given. Unless the phrase is followed by the word "ident," the pilot is not expected to ident after changing channels.

Stand by: Keep tuned to this frequency. I'll be right back.

Stand by one: Same as above, implying that the delay will be short, like one moment, minute or such.

That is correct. Self explanatory.

Verify: Please confirm.

Wilco: I have received the information or clearance and will comply.

Word twice: When used as a request it means that because of reception difficulties the person is asked to say every word twice when repeating the message. If used by the person transmitting a message, it means that because of apparent communications difficulties he will say each word twice.

In addition there are accepted ways in which to pronounce the words which represent the phonetic alphabet, and all numbers from zero through nine:

Alpha: Al-fah
Bravo: Brah-voh
Charlie: Char-lee or shar-lee
Delta: Dell-tah
Echo: Eck-oh
Foxtrot: Foks-trot
Golf: Golf
Hotel: Hoh-tell
India: In-dee-ah
Juliet: Jew-lee-ett

Kilo: Key-loh
Lima: Lee-mah
Mike: Mike
November: No-vem-ber
Oscar: Oss-cah
Papa: Pah-pah (often used as pop)
Quebec: Keh-beck
Romeo: Row-me-oh
Sierra: See-air-rah
Tango: Tang-go
Uniform: You-nee-form or oo-nee-form
Victor: Vik-tah
Whiskey: Wiss-key
Xray: Ecks-ray
Yankee: Yang-key
Zulu: Zoo-loo

1: Wun
2: Too
3: Tree
4: Fow-er
5: Fife
6: Six
7: Sev-en
8: Ait
9: Nin-er
0: Zero

I might point out here that, even though controllers in foreign countries communicate with local traffic in the local language, the official aviation language in all non-communist countries is English (virtually all air/traffic controllers are trained by the FAA in Oklahoma City), and even though occasional heavy accents may make communication difficult, it does not require the knowledge of another language to fly outside the United States. What is needed is a knowledge of the regulations which apply to the use of the airspace of that particular country.

Navigation

While there is little difference between communication radios except possibly for outward appearance and the quality of the components, navigation radios come in a great variety of types, each type designed to perform one or several different functions.

Nav Receiver

The most popular and widely used is the basic VHF *nav receiver*. It is a radio receiver which is designed to receive radio signals beamed by ground-based navigation aids, known as VORs (very high frequency omni directional radio ranges). When such signals are received they activate a cockpit display, known as the *omni-bearing indicator* (OBI) which tells the pilot on what radial from or bearing to that VOR he is currently located. It does not tell him the position of the aircraft's nose relative to that station, nor does it tell him how far he is from the station except that he is within reception distance. While the function of a VHF nav receiver and the associated OBI is really quite simple, it seems incredibly difficult to explain to a pilot who has no experience with the system. For this reason, there will be a detailed explanation with adequate illustrations in the chapter on navigation.

In addition to accepting and translating the above described radio signals from VORs, nav receivers are also equipped to receive voice transmission from the same station over the same frequency. For this reason standard weather broadcasts, the so-called sequence reports, are broadcast on these VOR frequencies, and pilots, if they so desire, can contact VORs on one of the published receive-only or simplex frequencies and request that the reply be transmitted over the VOR nav frequency.

Navigation frequencies are in the same VHF wave band as the communication frequencies, but their range is from 108.0 to 117.95 MHz, with the odd tenths frequencies from 108.1 to 111.9 being reserved for ILSs (instrument landing systems).

As was mentioned with reference to communication radios, many of the newer nav receivers are capable of storing one or more frequencies in addition to the one currently in use. Also, on most aircraft, two nav receivers will be installed (Fig. 12-3). In this instance this is not only for the purpose of reliability resulting from redundancy, but also because it requires reception from at least two VORs in order to determine the present position of an aircraft. Since VOR signals do not provide distance-to-the-station information, the only way to determine the aircraft position with a degree of precision is to tune to two different VORs, either in quick succession or, if two receivers are available, using one for each. Then, by a process of triangulation, the position of the aircraft can be reliably determined.

We might mention here that several manufacturers produce something called a *navcom*. As the name implies, these are

Fig. 12-3. A nav receiver incorporating an OBI in one box (courtesy of Narco).

systems which combined communication and navigation capabilities in one box. Even though they may share some of the working components, they work, for all practical purposes, like separate com and nav radios.

ADF

The *automatic direction finder* or ADF is the oldest electronic navigation aid used in aviation. It operates in the low-frequency wave band, being able to respond to signals being broadcast on frequencies from 190 kHz to 1799 kHz (kiloHertz or kilo-cycles).

An ADF consists of a tuner, a cockpit readout, and voice-reception capability (Fig. 12-4). When it is tuned to a transmitting station, the needle in the cockpit readout will point to that station, showing the direction to which the nose of the aircraft must be turned in order to fly to the station. Depending on the time of day and the atmospheric conditions, the ADF is capable of receiving signals from much greater distances than receivers operating in the VHF wave band because, unlike VHF waves, the LF waves do not propagate along the line of sight, but bounce back and forth between the surface of the earth and the base of the ionosphere. It is not at all unusual to pick up a station a thousand or more miles distant. For this reason it is always extremely important to listen to the transmission from the station in order to make sure that the one we are tuned to is actually the one we want to be tuned to.

For a variety of reasons which need not to be explained here in detail, ADF reception tends to become unreliable during dusk and dawn, and during periods of major atmospheric disturbances. And, when there are thunderstorms in the vicinity, the needle tends to point to the location of intensive lightning activity.

With VHF navigation equipment being easier to use and much more interference-free, the ADF is rarely used these days for basic navigation within the United States. Most of the time it simply serves as a back-up system. Beyond the borders of our country, on the other hand, we often find that VHF nav aids are quite scarce, and here, being able to use the ADF and to understand its functions can become important. In addition there are a number of airports in the U.S. which have published ADF instrument approaches and instrument pilots, in order to be able to use these approaches, must have an ADF on board.

The primary sources for reception by an ADF are the so-called non-directional beacons (NDB), low-frequency navigation aids which broadcast a Morse-code identifier, sometimes

Fig. 12-4. An ADF frequency-control unit (courtesy of Genave).

supplemented by a voice identifier, and some transmit continuous weather information; and standard broadcast stations. Thus an ADF can be tuned to any on-the-air radio station and the needle will point toward that station. When using such standard broadcast stations for navigation, it is often difficult to be sure that we are actually tuned to the right station, because radio stations don't identify themselves with any degree of frequency, and it may require listening to a whole bunch of silly commercials in order to figure out the town in which the station is located.

Most older ADFs use a coffee-grinder-type handle or a turning knob for station selection while the newer types are crystal controlled, making precise frequency selection much easier. Also, the older ones have fixed compass roses behind the needle, the 0 position of which is always at the top. With these units, flying a given direction with relation to a station can get to be a fairly exasperating experience. Newer models have a rotatable compass rose (Fig. 12-5). With this, the pilot can put his current direction of flight under the nose of the aircraft (at the top of the dial) and he is

Fig. 12-5. An ADF readout with rotatable compass card (courtesy of King).

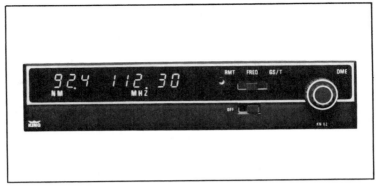

Fig. 12-6. A digital DME (courtesy of King).

then easily able to determine the direction of the station from his current position.

Using ADF transmitting stations for navigation is relatively simple when we fly toward the station. It becomes quite complicated if, for one reason or another, we are forced to navigate relative to a station which is behind us. The reason is simple. No matter in which direction we fly away from the station, the needle will always point to the tail of the aircraft as long as the station is directly behind us. More on this in the chapter on electronic navigation.

DME

Distance-measuring equipment or DME falls into the category of so-called pulse systems (Fig. 12-6). It sends out interrogating radio transmissions at regular (very short) intervals (pulses) which are received by appropriately equipped ground-based nav aids (VOR/DMEs and VORTACs) which send out a reply to each pulse. By measuring the elapsed time between the transmission of the interrogation and the reception of the reply, the system computes the distance of the aircraft from that particular nav aid. By using this information the cockpit display is capable of showing several types of data: the distance to (or from) the station in nautical miles; the current ground speed; and some will also show the time to the station assuming no change in ground speed.

The information provided by DMEs is usually reliable down to a fraction of a mile as long as the aircraft is a fair distance from the station. Once it gets close, there is an error, the result of what is known as the slant range. The easiest way to explain this is when visualizing an airplane at 10,000 feet above the elevation of the nav aid to which the DME is tuned. While the airplane is directly over the station, the DME will read just under two miles, because the airplane is, in fact, two miles vertically from the station. Some very sophisticated (and expensive) DMEs are equipped with automatic slant-range correction which eliminates this error, but most such systems are beyond the means of pilots flying the average general-aviation aircraft.

DMEs are not cheap, but they are a truly marvelous convenience item. No longer do we need to triangulate between two VORs in order to determine where we are, and no longer do we have to check the time it takes to get from one checkpoint to another. All this information is instantly available by checking the numbers which appear in the DME readout window.

In addition to all this, a DME is an integral part of the so-called area-navigation systems which have been rapidly gaining in popularity.

RNAV

The term *area navigation* (RNAV) refers to the ability to navigate with precision on routes which are not defined by the airways. This is accomplished by one of these minor miracles of which the current state of the art in electronics is capable. This miracle is the ability to electronically "move" an existing ground-based VOR/DME or VORTAC (it doesn't work with simple VORs) to any location at which we wish such an nav aid were actually located.

The piece of equipment which accomplishes this minor miracle is known as the navigation computer (Fig. 12-7). The pilot "tells" the computer on which radial or bearing from a given VORTAC, and at what distance from that VORTAC he would like his phantom nav aid, known as a *waypoint*, to be located. The computer, using information obtained from the nav receiver and the DME, then performs the necessary mathematical computations and positions the waypoint at the desired location. Once that has been accomplished the pilot can fly to or from the waypoint just exactly as if it were an actual nav aid, and the DME will provide the distance and ground-speed information with reference to the waypoint instead of the actual VORTAC.

The navigation potential and advantages of RNAV are so numerous and varied that I have written an entire book on the subject. Entitled *Practical Area Navigation* (TAB Book #2286), it

Fig. 12-7. The control-display unit of a sophisticated area-navigation system (courtesy of Collins).

explains these systems in considerable detail, more detail than we can go into within the limitations of this book.

Transponder

A *transponder* is a necessary piece of equipment if we want to operate in today's airspace without inconvenient restrictions. But it is a piece of equipment which aids the pilot only indirectly (Fig. 12-8).

The transponder, too, is pulse equipment. It sends out pulses which are received by radar stations and which then generate a clearly visible radar return on the radar scope of the air-traffic-control facility. Radar, of course, is capable of picking up aircraft in flight, even if they are not transponder equipped. But these returns are often weak, and in order to determine the particular aircraft which is generating the return, the controller must ask the pilot to fly circles or certain patterns. The transponder increases the visibility of the radar return and, by asking the pilot to "ident" he can instantly determine that the blip he is looking at actually

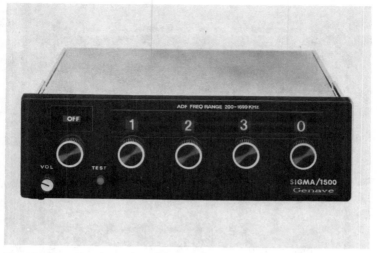

Fig. 12-8. Transponder (courtesy of Narco).

represents the aircraft he is talking to. Identing, meaning to press a little button on the transponder, results in a momentary drastic brightening of the radar return, thus identifying the aircraft in question.

Transponders are primarily a convenience item for the controller. But in today's airspace, with practically all aircraft being transponder equipped, a non-transponder-equipped airplane, if it should need help from air-traffic control, may find itself all alone out there and left to its own devices. This doesn't mean that emergency assistance is not available to a pilot without a transponder. It simply takes more time and makes things more difficult all around.

Transponders are mandatory in order to operate legally in terminal control areas and at altitudes above 12,500 feet, thus further restricting the operation of any airplane that is not so equipped.

Encoding Altimeters

An *encoding altimeter* or altitude digitizer is an adjunct to the transponder. Radar can locate an airplane only with reference to its horizontal position; if information about altitude is needed by the controller, he has to ask the pilot and has to assume that the pilot is telling him the truth. Encoding altimeters are designed to tell the transponder continuously the altitude at which the aircraft is flying, and the transponder, in turn, transmits this information to the radar facility where it appears in the form of a digital data block on the controller's radar scope. Thus he knows at all times not only where the airplane is laterally, but also how high it is vertically.

Like transponders, encoding altimeters are becoming mandatory equipment in more and more portions of the airspace, and the time may not be too distant when flying without one will place unacceptable limitations on the operation of an airplane.

Chapter 13
VFR and IFR

There are two distinctly different ways of flying our airplane. VFR, meaning in accordance with what are known as *visual flight rules*, and IFR, which calls for adherence to *instrument flight rules*. Both have their advantages and disadvantages. Let's take a closer look.

VFR

The rules applicable to VFR flight are quite simple. Basically they are simply weather minimums in terms of visibility and/or ceiling which permit the pilot to keep out of the way of other airplanes by the see-and-be-seen principle. These rules involve three parameters: *visibility* expressed in terms of statute miles; *ceiling* in feet agl; and *distance from clouds*, also in feet. When flying at an altitude of 1,200 feet or less in uncontrolled airspace, the visibility minimum is one statute mile. There is no minimum ceiling requirement and, with reference to clouds, the pilot is simply required to stay clear of clouds. When operating at an altitude between 1,200 feet agl and 10,000 feet msl in uncontrolled airspace, the visibility requirement calls for one statute mile, and the pilot is required to stay either 500 feet below the base of clouds, 1,000 feet above the tops of clouds, or 2,000 feet horizontally from clouds.

When operating in controlled airspace at or below 1,200 feet agl, the visibility minimum is three miles and the required distance from clouds is the same given above. When flying between 1,200 feet agl and 10,000 feet msl in controlled airspace the rules remain the same as those in effect below 1,200 feet. At altitudes above 10,000 feet, except when within 1,200 feet of the terrain in mountainous areas, the visibility minimum is increased to five statue miles and the distance-from-clouds requirements call for 1,000 feet either above or below, and one mile horizontally, regardless of whether the airspace is controlled or uncontrolled.

In addition, when operating to or from an airport with an operating control tower, or when flying through the airport traffic area of such an airport, the ceiling must be not less than 1,000 feet, and the ground visibility must be reported as three miles or better. (For an exception to this rule, see **Special VFR** below.)

The only other rule involving VFR operations concerns the flight altitudes which VFR aircraft are expected to maintain in cruise. At altitudes below 3,000 feet agl there are no specific requirements. Any flight level will do, no matter the direction of flight. At altitudes above 3,000 feet agl and up to 29,000 feet msl, VFR pilots flying a course (not heading) anywhere between zero and 179 degrees (magnetic) are expected to fly at odd-thousands-plus-500-foot altitudes, meaning 3,500 feet, 5,500 and so on. These altitudes are expressed in feet msl (not agl). When operating on a course between 180 degrees and 359 degrees (magnetic) the cruising altitudes are even-thousands-plus-500 feet such as 4,500, 6,500 feet msl and so on. Above 29,000 feet msl, an altitude where VFR flight is rarely conducted, the altitudes for aircraft flying toward the east (zero to 179 degrees magnetic) are flight levels at 4,000-foot intervals starting at 30,000 feet, such as 30,000, 34,000, 38,000 and so on (usually expressed in terms of flight levels: FL300, FL340, F380). And for westbound flights (180 to 359 degrees magnetic) the 4,000-foot intervals start at 32,000 feet, such as 32,000, 36,000, 40,000 (or FL320, FL360, FL4000).

In practice, the only rule which really concerns the average pilot is the 1,000-foot ceiling and three mile visibility restriction at controlled airports. All that other stuff about visibility in flight and distances from clouds is something which keeps showing up in the FAA written examinations, but something which means very little once we get into an airplane and up in the air. The fact is that it is completely impossible to judge with any degree of accuracy whether the visibility during haze conditions is one mile, three miles, five miles or whatever. The only time we have a chance to get some kind of a rough idea of the flight visibility is when there is another aircraft more or less at our altitude. But since that doesn't happen very often, our only logical basis for gauging the visibility is whether or not we feel reasonably comfortable with what we can and cannot see.

Similarly, I defy anyone to accurately judge the distance of the aircraft from clouds, regardless of whether that distance is horizontal or vertical. Clouds have no sharply defined lines and no two of them are ever of the same size. Thus, to know with any degree of certainty whether that distance is the required 1,000 or 2,000 feet, is something no one I know has ever been able to do.

To put matters succinctly, a VFR pilot can take off from an uncontrolled airport if the visibility is one mile regardless of ceiling conditions, assuming that he will be able to stay clear of clouds. He can take off from a controlled airport if the visibility is three miles and the ceiling (meaning clouds covering 60 percent or more of the sky) is not less than 1,000 feet. Once he is airborne and beyond the limits of the airport traffic area, the ceiling limitation no longer applies.

During the en-route portion of a VFR flight the pilot is not required to talk to anyone on the ground or to make use of any of the available navigation aids. If he so desires he can operate strictly by pilotage, meaning that he looks at landmarks on the ground and uses them as a guide to get him to where he wants to go. Or he can use dead reckoning, meaning that he simply takes up the heading and holds that heading until he arrives at his destination. Or he can tune in his navigation radios and utilize all or any number of the ground-based nav aids for the purpose of finding his way, and he *still* doesn't have to talk to anyone.

On the other hand, it is perfectly legitimate for the VFR pilot to contact any of the available FAA facilities (with the exception of ARTCCs, which usually handle only IFR traffic. They will assist a VFR pilot in trouble, but that is the exception rather than the rule), and ask for weather information or any other assistance which he may feel he needs.

In this manner, the VFR pilot can fly for thousands of miles, left entirely to his own devices with no one telling him what to do, using whatever route and altitude appeals to him, as long as he stays out of positive control airspace, such as TCAs, and flies around or stays above airport traffic areas. He can fly below the clouds or climb up high and operate VFR on top of the clouds, and he can do this at any time of day or night. There is no other country in the world in which the VFR pilot can operate with such a degree of complete freedom, and this continues to be true despite the fact that the FAA keeps on devising restrictions which did not exist in the past.

Assuming a degree of savvy with reference to weather, the VFR pilot will find that most of the time he can take off and get to his destination with little difficulty and few delays. There are times, of course, when VFR flight is impossible, or when conditions are so marginal as to make continued VFR flight less than safe. It is entirely up to the pilot to know when to fly and when conditions are such that, based on his level of experience and the type of aircraft and the equipment on board, a flight should be postponed or abandoned.

Special VFR

Special VFR, often expressed in writing as S/VFR, is a rule which was devised to make it possible for pilots to take off from or land at controlled airports when weather conditions are less than the standard VFR minimums. The minimums for S/VFR are the same as for uncontrolled airports, namely one mile visibility and the ability of the pilot to remain clear of clouds. There is no ceiling restriction associated with S/VFR.

At some busy terminal airports S/VFR operations are not permitted. These airports are identified on Sectionals by control zones outlined with a line of Ts (TTTTTTTTTTT) in the same blue color which is used for the usual designation of control zones (--------). This rule is strictly enforced and the only way to take off from or land at such an airport when conditions are below VFR minimums is to file IFR (Fig. 13-1).

At all other controlled airports, the tower controller must permit a VFR pilot to takeoff or land in accordance with S/VFR, assuming that the request for S/VFR originates with the pilot. Tower controllers will *never* suggest S/VFR to a pilot. When an S/VFR operation has been requested it must defer to any IFR traffic operating within the applicable airspace. This means that the VFR pilot who wants to take off with an S/VFR clearance, or who wants to land that way, may be held either on the ground or in VFR conditions in the air for quite a while until there is no IFR traffic in the area.

An S/VFR clearance consists of a certain, usually quite simple, set of instructions which includes the direction of flight after takeoff, an admonition to stay clear of clouds, and the request to inform the tower when VFR conditions on top have been reached, or when the aircraft is clear of the control zone. (In this case it is the control zone rather than the airport traffic area which is of consequence, because, technically, S/VFR is considered an IFR operation.) For an S/VFR landing the instructions are quite similar. Usually the pilot will be told to stay clear of clouds while approaching and then entering downwind (or base), at which point he is expected to have the airport in sight, permitting a normal visual landing. In either case the pilot is expected (though not required unless asked by the tower) to read back the clearance he was given in order to make sure that it has been understood correctly.

In practice, no VFR pilot should attempt to make use of the S/VFR option unless he is capable of controlling his aircraft by instruments alone. Despite the always repeated admonition to stay clear of clouds, there is always the possibility of finding oneself in a cloud, if only for a brief moment. To the inexperienced pilot this can be quite a shock, and he must be able to remain calm, composed, and in control of his airplane for whatever time it takes to get back into VFR conditions.

Some pilots, though they may be instrument rated (or, at least, instrument proficient), prefer to fly VFR whenever possible. They like the freedom of choice, the fact that there is no one to tell them where to fly and what to do. They also believe, with a degree of justification, that operating VFR, flying the most direct route from point to point, is more economical in terms of time as well as fuel. Others like the instrument environment and they will, regardless of the weather, file instruments each time they take off. Except when the weather is strictly IFR, this is simply a matter of personal preference. Neither can say that his way is better or more right than that of the other.

IFR

Flying IFR is the exact opposite of flying VFR. When on instrument flight plans, we are required to be in constant contact with persons on the ground, to tell them what we intend to do before we do it, and to obtain a clearance for each flight and any deviation from the original flight plan. All IFR operations are conducted under the control of ATC (air-traffic control), and ATC's sole purpose in exercising this control is to provide safe separation between cooperating aircraft. The word cooperating is important here. ATC *cannot* and *does not* provide separation from aircraft operating VFR. Thus, except when actually in clouds and thus

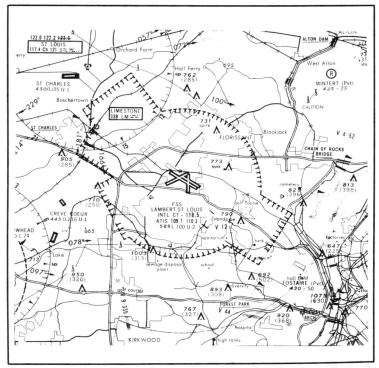

Fig. 13-1. A control zone in which special VFR operations are not authorized, as shown on Sectional charts.

totally deprived of all visibility, the IFR pilot must maintain a constant visual watch for other aircraft.

While filing a flight plan is a voluntary and relatively meaningless exercise for the VFR pilot (its only purpose is to automatically instigate a search if the pilot fails to arrive at his destination), the instrument pilot *must* file a detailed flight plan giving his route of flight, altitudes and so on, with ATC and he cannot start his flight until a clearance has been received from ATC. Such a clearance may be "as filed," meaning that ATC has accepted the pilot's requests for routing and altitudes, or it may be entirely different if the requested routes or altitudes represent some type of potential conflict with other IFR traffic.

At first glance it would seem that by filing instruments, the pilot relinquishes his command responsibility to ATC and the controllers handling the flight. This is *not* the case. ATC clearances, though often referred to as instructions, are *requests* rather

187

than *commands*. Any pilot who feels that for whatever reason he would prefer not to comply with an ATC clearance, has a perfect right to respond "Negative!" and to request an alternate clearance. The only restriction involved is that, except in case of an emergency, he must not deviate from the last clearance received until an amended clearance has been transmitted to him.

IFR flight effectively eliminates most of the weather minimums which tend to occasionally inhibit VFR flying.

For non-commercial flights (flights which carry neither passengers nor freight *for hire*), there are no takeoff minimums at all. Such a flight may completely legally take off under zero-zero conditions when all the pilot can see are maybe 20 or 30 feet of runway in front of him. It may not be a *smart* thing to do, but it's quite *legal*. Commercial flights, flights which carry either persons or freight for remuneration, *do* have to adhere to certain minimum visibility requirements to be able to take off legally, but these minimums are so low at most airports that the instances of IFR flights being delayed because of them are rare indeed.

While en route IFR flights need pay no attention to clouds (except thunderstorms. But the avoidance of thunderstorms is not for the benefit of ATC but solely for the safety of the flight). As far as the IFR pilot is concerned, clouds simply do not exist, except with reference to the fact that while inside a cloud the pilot is relieved of the need to look out for VFR aircraft, if only for the reason that he wouldn't be able to see another aircraft until it was within a few feet of him.

IFR flights normally conclude with an instrument approach to a landing at an airport which has one or more published instrument-approach procedures. Instrument approaches are divided into *precision* and *non-precision* approaches. Precision approaches are those which include vertical guidance in the form of a glide slope or in the form of the (today rarely used) vertical-guidance capability of precision radar. Non-precision approaches are all the other types, all of which are based on horizontal guidance and the use of time and prescribed rates of descent to bring the pilot to that point of the approach (the missed-approach point or MAP) where he must have the airport in sight in order to be able to execute a visual landing. To date, although the necessary equipment is available, no landings under zero-zero conditions are approved for civil aircraft in the United States.

There are literally thousands of such instrument approach procedures, all of which are published (both by the Government

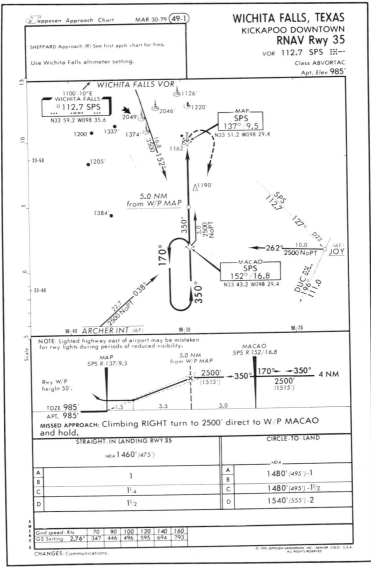

Fig. 13-2. A typical chart for an instrument approach (copyright by Jeppesen Sanderson, Inc.).

and by Jeppesen) on so-called approach plates or charts (Fig. 13-2). These approach plates describe in words as well as in graphic representations the procedure which must be flown in order to get the aircraft safely to the MAP. Since there are nearly as many different procedures as there are instrument approaches, no pilot could be expected to remember these precedures, and no such approach may be legally flown unless the appropriate approach plate is on board the aircraft.

For reasons having to do with the terrain and other obstacles in the vicinity of the airport and the degree of complication of the approach procedure, the landing minimums vary greatly. All these minimums are expressed in terms of visibility, either in statute miles or fractions thereof, or in terms of feet of runway visibility range (RVR). While there is never a published ceiling limit, there is one in practice, because each such approach includes a minimum descent altitude (MDA) in the case of non-precision approaches, and decision height (DH) in the case of precision approaches. When the MDA or DH has been reached and the ground is not in sight, the aircraft may not descend any further. It may continue at that level to the MAP and if there is still no satisfactory ground contact, a missed approach must be flown. Thus, the ceiling minimum is, for all practical purposes, the MDA or DH (though in the case of a broken overcast which technically represents a ceiling, an approach may be continued to the landing, even though the bases of the nearby broken clouds are below the MDA or DH).

While it would go beyond the scope of this book to give a complete set of instructions of how to fly all types of instrument approaches, here is a quick analysis of the different categories.

Precision approaches are either approaches using a complete instrument landing system (ILS) or precision approach radar (PAR). An ILS consists of a localizer for horizontal guidance and a glide slope for vertical guidance, both providing the pilot with information with reference to his position through a cockpit readout instrument (OBI). In addition there is a compass locator beacon (LOC) and an outer, middle and occasionally inner marker (OM, MM, IM) to tell the pilot how far along he is on the approach. All the pilot has to do is to keep two needles in his OBI, one horizontal and one vertical, centered, and he is assured of safe obstacle clearance and can depend on the fact that when he breaks out of the clouds, the runway will be right in front of him.

PAR approaches depend on a controller on the ground telling the pilot which way to fly and at what rate to descend. They are virtually non-existent in today's civil aviation and, with few exceptions, can only be flown at military airports.

Non-precision approaches are all of the following: Localizer approaches, back-course approaches, VOR approaches, ADF approaches, surveillance-approach-radar approaches, and circling approaches. In all except the last case, the name of the approach has to do with the type of navigation aid to be used. Localizer approaches and back-course approaches use either the front side or the back side of a localizer without a glide slope. This may be either because there is no glide-slope instrumentation either on the ground or in the aircraft, or because the glide slope is temporarily out of service.

VOR approaches and some referred to as VOR/DME approaches use a VOR or VORTAC for navigational guidance, while VOR/DME approaches combine that guidance with the information available from an on-board DME.

ADF approaches use either a non-directional beacon (NDB) or, in some cases, a standard broadcast station for navigational guidance.

The surveillance-radar approach (SAR or GCA), also often referred to as a ground-controlled approach, uses a radar which is incapable of determining the altitude of the aircraft. A controller watching his radar scope will give vectors to the pilot and tell him at what altitude he should be. It is up to the pilot to comply with those verbal instructions.

Circling approaches are really portions of non-precision approaches under certain weather conditions. Each published approach is designed to bring the pilot to a point from which he can make a straight-in (or nearly straight-in) approach to touchdown at a given runway. When wind conditions are such that landing on that runway is impossible or, at least, not advisable, the so-called circling maneuver is executed once the pilot has the airport in sight to permit him to line himself up with the active runway. Circling approaches can be quite ticklish and they are supposed to be flown within the published guidelines and rigid restrictions.

Obtaining an instrument rating is much like getting pregnant. There is no such thing as a little bit of it. Even though you may never want to fly an instrument approach down to minimums, you must be qualified to do so in order to obtain the rating. This involves considerable study and practice, the passing of a flight test and of a rather difficult written examination (plus a minimum of 250 hours as pilot in command). It's expensive and time consuming, but to those who want to get maximum utility out of their aircraft, it may be a necessity.

While the VFR pilot can get away with a degree of sloppiness, instrument flying calls for great precision. Lateral deviations from the airway centerline are frowned upon unless specifically authorized by ATC. Altitude must be maintained with care because there may be aircraft below or above which cannot be seen because of the clouds and the resulting restrictions to visibility. Climbs and descents must be accomplished in accordance with ATC clearances, as those instructions are based on the operations of other aircraft which may need the airspace which, moments ago, was vacated by the pilot. If for one reason or another compliance with a clearance is either impossible or contrary to the pilot's judgement (as often happens during conditions of severe icing in clouds) ATC must be so informed and an amended clearance must be requested.

IFR flying is a strictly structured system designed to keep airplanes out of one another's way. It works reasonably well most of the time, but it does away with the freedom of the skies of the VFR pilot.

Chapter 14
Electronic Navigation

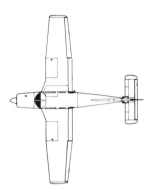

Since by far the greatest amount of our navigation these days is dependent on the use of electronic gimmickry, it seems appropriate to devote a separate chapter to the subject. The first thing that a pilot should learn to understand, even before trying to learn how to use these electronic navigation aids, is the basic principle on which they operate. Once this is clearly understood, it becomes considerably less confusing to try and interpret the signals which are presented in one form or another in the cockpit.

The Principles of VORs

A VOR (the term stands for very high frequency Omni-directional radio range) is a station which sends out an unlimited number of signals in every conceivable direction of the compass. To simplify things, let's just say that it sends out 360 beams, one along each degree of the compass rose. How this is done, and how one beam differentiates from the other is of no importance or interest to the pilot. All he needs to know is that his VHF nav receiver is capable of receiving these beams, of determining which individual beam is being received at a given moment, and, by way of a cockpit display, to tell which beam it is he is currently receiving.

In order to utilize the VOR network for navigation it requires therefore the interaction of three instruments: the VOR station on the ground; the airborne nav receiver; and the cockpit display. The station on the ground is simply there, and its location is shown on all aviation charts (Fig. 14-1). It operates 24 hours a day (except when down for maintenance) without the help from any human operator. It operates on a given navigation frequency which is also listed on all aviation charts, and any pilot can tune his receiver to that frequency and receive the desired information without the need to talk to anyone. (In fact, VORs are also capable of transmitting voice signals on that same frequency which can be received by the pilot at will by simply turning up the volume on his nav receiver.)

The airborne nav receiver is a radio which can receive but not transmit radio signals. It is tied in to the loudspeaker and/or earphones for the purpose of projecting voice and other audio signals to the pilot or into the cockpit. It is also tied to a visual cockpit display which transforms the radio signals into a meaning-

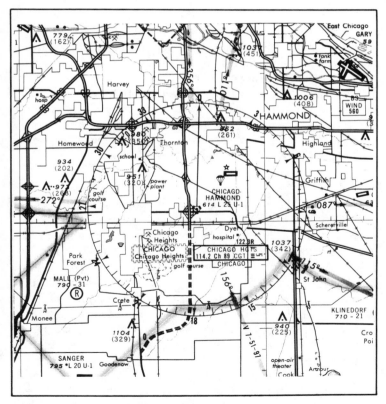

Fig. 14-1. A VOR as shown on Sectional charts.

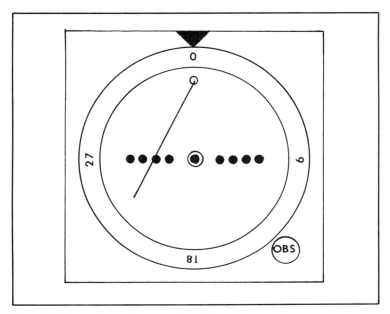

Fig. 14-2. A typical face of a conventional OBI.

ful readout. The nav receiver itself is used only for frequency selection and volume control. Without the cockpit display it is worthless except for the purpose of receiving voice or other audio transmissions.

OBI

The cockpit display is commonly referred to as the omni-bearing indicator, or OBI for short (Fig. 14-2). An OBI consists of a course-deviation indicator or CDI, and an omni-bearing selector or OBS. Since this is the instrument which must be thoroughly understood by the pilot, and since different manufacturers have produced different types of OBIs, we will now spend a little time to explain exactly how this rather amazing instrument works.

The OBS is a turning knob which operates a rotating compass rose (Fig. 14-3). The pilot can and must use it to place a selected VOR radial or bearing under the dot or arrow which indicates that this is the radial or bearing with reference to which all indications provided by the OBI must be interpreted. In other words, if the OBS is set to 360 degrees, all indications provided by the OBI will show the aircraft's position with reference to the 360-degree radial from or bearing to the VOR to which the nav receiver is tuned. If the OBS is now moved to a different radial or bearing, the OBI display will change to new information relative to that new radial or bearing.

Let's clear up a bit of confusion which usually exists because the terms *radial* and *bearing* are often used interchangeably and, therefore incorrectly. A radial is the line running *outward* from the center of the VOR along a given degree. The bearing is the line running *inward* toward the center of the VOR along a certain degree.

Here is how this works: We tune the nav receiver to the frequency of a certain VOR, one which, according to our estimated position and the charts, is within reception distance. Since VORs transmit on the VHF waveband, and since VHF waves emanate from the source in a straight line (so-called line-of-sight), they can only be received within limited distances, distances which increase with the aircraft's altitude above the ground. Thus the average reception distance, assuming no mountains are in the way, is as follows:

Altitude agl in feet	Line of sight in statute miles
500	30
1,000	45
2,000	65
3,000	80
5,000	100
7,000	120
10,000	140
15,000	175

To make certain that the nav receiver is receiving signals from the VOR which we are interested in, we should now turn up the volume and listen to the station identifier which will consist either of a Morse-code signal representing the three-letter call sign of the VOR, or a voice identifying the VOR, or both. Once we're sure that we are receiving the right station, we now turn our attention to the OBI. We twist the OBS knob until the CDI indicator reacts, and we then fine-tune it until the CDI needle (or its equivalent) centers. Let's say, for argument's sake the needle centers when the OBS is tuned to 035 degrees. We now know one of two things. We're either positioned somewhere along the 035 radial or along the 035 bearing with reference to the VOR. But which? To determine that we look at the one additional indication provided by the VOR which

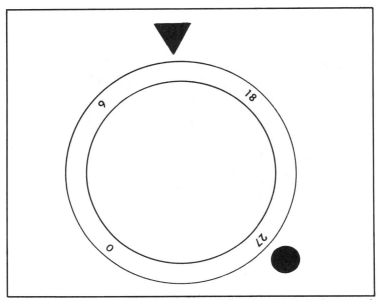

Fig. 14-3. The omni-bearing-selector knob (OBS) at the right bottom corner of the instrument is used to align the compass card with the indicator at the top.

so far has not been mentioned. This is the TO-FROM indicator, also referred to as the ambiguity meter. It consists of a small window in the face of the OBI which will display either the letters TO or FR (or T or F) or an arrow pointing either up or down, depending on the type of instrument. If, in this example, the displayed letters read TO (or the arrow points up), it means that we are reading the direction *to* the VOR. This places the position of the aircraft on the 035-degree *bearing* to the VOR which is also the 215-degree *radial* from the VOR. If we would now twist the OBS knob again and line it up with 215 degrees, the TO-FROM display will change to FR (or the arrow will point down), telling us that we are currently located on the 215-degree radial from the VOR (Figs. 14-4 through 14-9).

So far we have found how to determine the aircraft's position with reference to the *direction* to or from the VOR. What we do not know is the *distance* from the VOR, nor is the reading influenced in any way by our direction of flight. There is no easy way to determine the distance from the VOR using just one such nav aid (unless the aircraft is equipped with DME). In order to determine the distance and thus the exact position of the aircraft, we now need to tune the nav receiver (or, if two such receivers are on board, the

Fig. 14-4. The course-deviation indicator (CDI) is the needle in the center of the instrument. The one at right is a simple VOR receiver while the one on the left includes a glide slope receiver (courtesy of Narco).

second nav receiver) to a second VOR which must also be within reception distance. We go through the same routine described above. This time, let us say that the needle centers when the OBS is set to 325 degrees, and the TO-FROM indicator reads TO. Now, if we draw a line through the two VORs (either mentally or on our

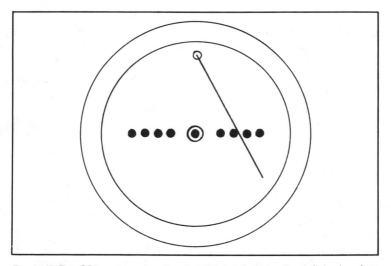

Fig. 14-5. The CDI centers when the aircraft is located atop the radial or bearing aligned with the top mark of the OBS. In the instance shown here the aircraft is to the left of the selected radial or bearing, and the pilot must fly toward the needle to intercept it.

Fig. 14-6. Instead of using a swinging needle or moving bar as the CDI, Bendix has developed an OBI using lighted bars. In this instance the aircraft is to the right of the 125° radial or bearing of the selected VOR and the pilot must correct to the left to intercept it.

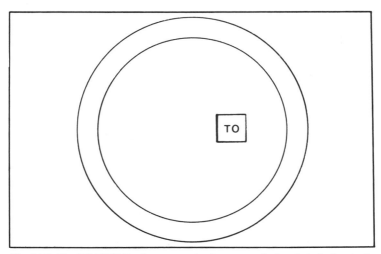

Fig. 14-7. The TO-FROM indicator or amiguity meter tells the pilot whether he is navigating with reference to the radial from a VOR or the bearing to a VOR.

Fig. 14-8. Some OBIs use moving bars as CDIs and some have arrow heads in place of the TO/FROM indication (courtesy of King).

chart) which coincides with the two readings we have received, these lines will cross somewhere (Fig. 14-10). The point at which they cross represents the current position of the aircraft (give or take a mile or so, since all this takes a few minutes during which the aircraft continues to move). By now measuring the distance from the point at which the two lines cross to either of the two VORs, we can determine the distance to that VOR.

All this sounds a lot more complicated than it actually is. In practice we find that most of the time we'll be flying directly toward a VOR, staying on course simply by keeping the CDI needle centered. And most of the time we don't really care how far we are from the VOR. This is why, when a passenger asks the pilot, "Where are we?" he is likely to answer: "Somewhere along this line between here and there."

I said earlier that there is no *simple* way to determine the distance from a VOR by using only one VOR. There is a complicated one and frankly, in over 25 years of flying, I have never used it. But just to keep the record straight, here is how it works:

First we establish ourselves on a given radial or bearing. Then we turn the aircraft so that the VOR is lined up with the wingtip of the aircraft, in other words, that we are flying at right angles to that radial or bearing. We now check how long it takes to accomplish a 10-degree bearing (or radial) change (Fig. 14-11). We now multiply our true airspeed by the number of minutes it took to accomplish that 10-degree bearing (radial) change and divide the result by that amount of bearing change, namely 10. The resulting figure is the distance to the VOR. For example: If it takes two minutes to accomplish the 10-degree bearing change and if the true airspeed is 120 knots, we multiply 120 by two which produces 240. We now divide 240 by 10 and arrive at 24 which is the distance in nautical miles to the station. (If the true airspeed had been used in terms of mph, the result would have been in statute miles.) As I said, it's doubtful that you'll ever be tempted to use this procedure, but it

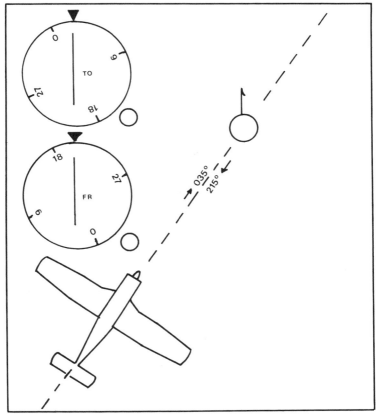

Fig. 14-9. The OBI readings which will appear when the TO or FR positions are selected.

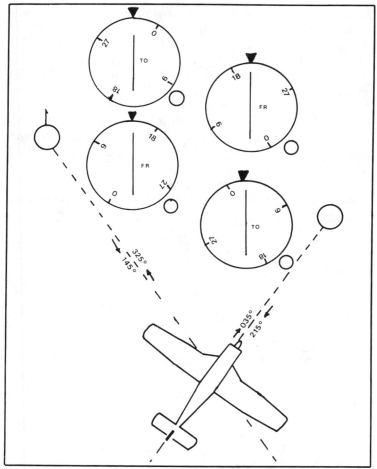

Fig. 14-10. By triangulating between two VORs, the pilot can easily establish the aircraft's current position.

could be helpful in an area where VORs are too far apart to receive more than one at a time.

Now a word about the direction of flight with reference to a VOR. It makes absolutely no difference which way the nose of the aircraft is pointing when it crosses or flies along a given radial or bearing. No matter the direction of flight, the OBI reading will always be the same (Fig. 14-12). Therefore, unless we're flying directly to or from a VOR, the readings are of no real value in keeping us on an off-airways course. (All Victor Airways are aligned in one way or another to a VOR) The only way in which

VORs (actually VOR/DMEs or VORTACs) can be used for precision navigation off airways direct is by using on-board area-navigation equipment.

Before getting off the subject of VORs it should be pointed out that the CDI-needle (or comparable indicator) deflection is angular. This means that a one-dot deflection represents a shorter distance from the centerline of the radial or bearing when we're close to the station than it does when we are a greater distance from the station. In fact, each one-dot needle deflection represents a two-degree deviation from the selected radial or bearing (Fig. 14-13). In the RNAV mode, meaning that the system is being used in conjunction with area-navigation equipment, the CDI-needle deflection is linear, meaning that each one-dot deflection represents .25 nm from the centerline in the LOC mode and one nm in the RNAV mode, regardless of the distance of the aircraft from the referencing station (See Fig. 14-14).

In practice, when flying in the United States, you will find that you'll do most of your navigating by using the VOR network. But there are occasions when using the ADF system of navigation is of advantage.

The Principle of ADF

The automatic direction finder (ADF) works on principles which are entirely different than those of the VORs. First of all the ADF uses different radio waves and these waves behave differently than do those in the VHF band. Second, the indicator associated with the ADF is not designed to give information with reference to a certain radial or bearing, but rather with reference to the position

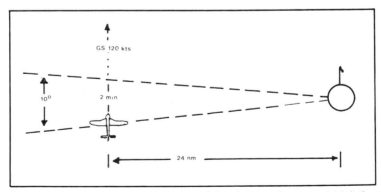

Fig. 14-11. By flying at right angles to a VOR radial, the pilot can establish the distance to the VOR, assuming he knows his ground speed.

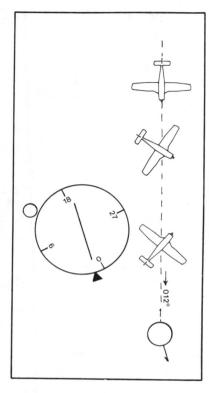

Fig. 14-12. There will be no change in the readout of the OBI as long as the aircraft is positioned above the selected radial or bearing, regardless of the direction of flight of the aircraft.

of the nose of the aircraft relative to the location of the ground-based station being used.

To start with, let's look at the behavior of those VLF (very low frequency) waves along which the ADF signals travel. These waves do not follow the line-of-sight principle. Rather, they bounce back and forth between the surface of the earth and the base of the ionosphere. They, therefore, can be thought of as following the curvature of the earth. As a result they can often be received for great distances, sometimes reaching thousands of miles. But they are temperamental. They tend to be disturbed by any number of atmospheric conditions. Without going into a lot of scientific goobledygook, ADF signals are unreliable during the hours of dusk and dawn, and in the general vicinity of thunderstorms. They are pretty good during the daytime and are usually best on clear nights. There is always a lot of static associated with the audio portion of ADF reception, which is the primary reason why virtually all voice communication in aviation uses the VHF frequency band.

The stations used by ADF fall into three basic categories: *non-directional beacons* (NDB) which are low-frequency navigation aids established specifically for use in aviation; *compass locators* and other components of localizers and ILSs; and *standard broadcast stations*, in other words, any normal AM radio station (not TV, which is FM). All that is required of any of these stations is that they are broadcasting continuously. As long as they are on the air the ADF will pick up their signals and produce a readout in the ADF cockpit display.

In the cockpit the pilot has a means of tuning his ADF to a given frequency. He can switch the system either to ADF which eliminates audio and simply activates the cockpit readout, or he can switch to a setting which permits him to receive and monitor whatever sound is being transmitted over the selected frequency. With the sound on the cockpit display is somewhat less reliable than it is with the selector set to ADF.

Since at times stations located at great distances can be received, it is always important to initially use the sound-reception

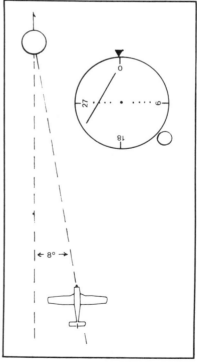

Fig. 14-13. The CDI deflection in an OBI tuned to a VOR is angular, meaning that a certain amount of deflection coincides with a certain number of degrees. Each dot indicates a two-degree deviation from the selected radial, but the actual distance will vary with distance from the VOR. In the illustration, if the aircraft continues to fly along the right dashed line toward the station, the needle will maintain the same amount of deflection until the station is reached.

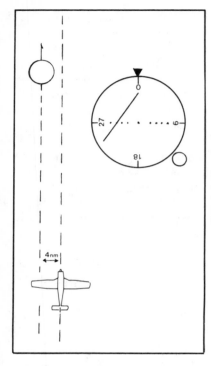

Fig. 14-14. When the nav receiver is used in conjunction with an area-navigation system, the CDI readout is linear, meaning that a one-dot deflection represents a distance of one nautical mile from the selected radial or bearing, no matter how far the airplane is from the station.

capability in order to make absolutely certain that the station being received is the one we are concerned with. NDBs and other low-frequency aviation nav aids continuously broadcast a Morse-code or voice station identifier. Not so standard broadcast stations. They identify themselves rather infrequently (they are required to do so only on the hour and the half-hour) and it may take a bit of careful listening to ascertain whom and what we are tuned to.

The cockpit display consists of a needle which, when it points straight up, means that the station we are tuned to is straight ahead of us. In other words, it looks at the direction of flight of an aircraft at any given moment and points *to* the transmitting station (Fig. 14-15).

To put it quite simply, the ADF indication is best and most easily used for the purpose of flying straight to a station. It can also be used in conjunction with signals from one VOR to achieve a triangulation for the purpose of determining the present position of the aircraft (Fig. 14-16). Though there are ways to use it effectively in flying away from a station in a given direction, this tends to become rather complicated and should not be attempted by

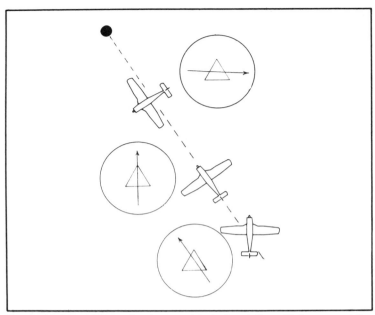

Fig. 14-15. The ADF needle always points to the station with the nose of the aircraft being thought of as being at the top of the instrument.

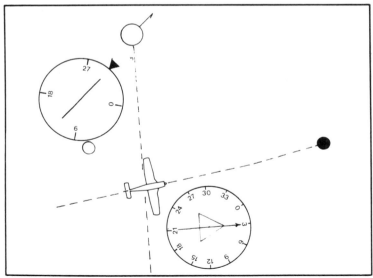

Fig. 14-16. A pilot can determine the position of his airplane by triangulation between the readouts from the ADF and the OBI, using one NDB or standard broadcast station with the ADF and a VOR with the OBI.

pilots who have not had a chance to accumulate a *considerable* amount of experience in working an ADF.

The problem with using ADF in flying away from a station is the result of the fact that as long as the tail of the airplane points straight at the station, no matter the direction of flight, it simply means that the station is in a straight line behind the airplane. But assuming an even minor crosswind component, this would eventually result in a curved flight which has no resemblance to the desired course. There are ways in which the experienced pilot can use the indications received by his ADF in flying a constant course in a given direction away from the station, but they are actually a bit too complicated to be explained in detail within the limits of this book. Figures 14-17 and 14-18 will explain what this is all about to those readers who might feel that this is important.

As a matter of fact, navigation by electronic aids tends to seem inordinately complicated when we try to explain it in written (or

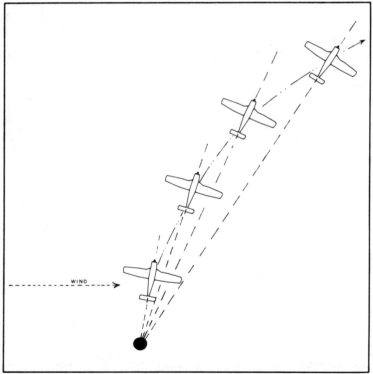

Fig. 14-17. Flying away from an NDB or a standard broadcast station will result in a curved flight unless adequate course corrections are made.

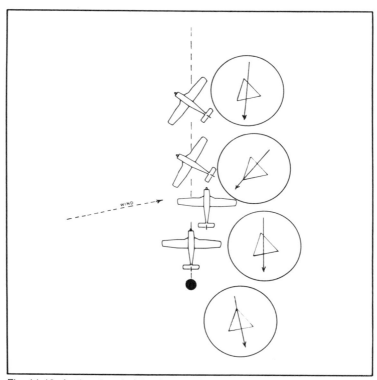

Fig. 14-18. As the aircraft drifts downwind of its course, a course correction is made. First this correction is quite extreme. It is then minimized to about half that correction to keep the needle a certain distance upwind of the straight-back position which will help to keep the aircraft on course.

spoken) words. A careful study of the different illustrations will tend to clarify the many ambiguities. What *is* important is that the pilot understands the basic principles of operation of VHF and ADF navigation, because all of this type of navigation requires that the pilot be able to form a mental picture of what is going on. The ability to create a mental picture of the aircraft's position with reference to stationary fixes is one of the most important prerequisites of flying without getting lost. In addition, this ability will be of immense help in the later stages of flying, when we decide to work on obtaining our instrument rating. Instrument flying is flying without reference to any visual clues other than those available in the cockpit, and being able to translate those clues into a three-dimensional visual image by instinct rather than a great deal of mental gyrations is the sign of a competent instrument pilot.

We can consciously work on this throughout all the years when all our flying is done VFR. Always try to have a mental picture of your airplane with reference to its ground-based surroundings. Eventually this will become second nature. When this happens to a pilot who has also learned to think of the angle of attack as the very basis for all flight, he then can feel justified in thinking of himself as a safe and competent pilot.

Glossary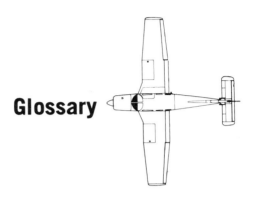

absolute ceiling—The maximum altitude above sea level to which a particular aircraft can climb and then maintain horizontal flight under standard atmospheric conditions.

accelerate-stop distance—The distance required to accelerate an aircraft from a standing start to liftoff speed, and, assuming failure of the (critical) engine at the instant that speed is reached, to bring the aircraft to a stop, using heavy braking.

ADF—Automatic direction finder.

aerodynamics—The forces, such as resistance, pressure, velocity and others involved in the movement of air or gases around a moving body. Conversely, the branch of dynamics and physics dealing with these forces.

agl—Above ground level.

AH—Artificial horizon.

ailerons—The primary control surfaces located at the trailing edges of the outer wing panels which, when moved up or down, cause the airplane in flight to bank.

air-data instruments—The basic instruments used by the pilot in the control of the aircraft, such as airspeed indicator, altimeter, turn-and-bank indicator and so on.

airfoil—Any surface designed to create lift, either positive or negative, when moving through the air at a given speed. Primarily wings and control surfaces, though propellers and helicopter blades are also airfoils.

airspace—When used in aviation the term means the navigable airspace, for all practical purposes between ground level and 60,000 feet.

airspeed—The speed with which an aircraft is moving with relation to the air around it. It may be expressed as *indicated* airspeed, *calibrated* airspeed and *true* airspeed.

airspeed indicator—A flight instrument with a cockpit readout which, in terms of knots or mph, shows the difference between pitot pressure and static pressure. The reading obtained from the airspeed indicator is indicated airspeed.

alternator—An electrical device, serving the same purpose as the old-style generator, which is driven by the engine and supplies current to the battery and to all on-board electrical equipment except the ignition system.

altimeter—A flight instrument capable of displaying the height above sea level (or any other predetermined level), activated by an aneroid barometer measuring atmospheric pressure at the given altitude.

altimeter setting—The barometric-pressure reading in the small window provided for that purpose on the face of the altimeter.

ambiguity meter—The TO/FROM indicator in an OBI.

angle of attack—The angle at which the chord line of the wing or any other airfoil meets the relative wind. It determines the amount of lift developed at a given airspeed.

approach—The maneuvers an airplane needs to perform prior to landing.

approach control—The ATC facility monitoring and directing traffic approaching an airport where such a facility is in operation.

ARTCC—Air Route Traffic Control Center, usually simply referred to as *center*. The ATC facilities handling en-route IFR traffic.

artificial horizon—A gyro instrument showing the attitude of the aircraft with reference to pitch and roll as compared to the horizon.

ATC—Air Traffic Control.

atmospheric pressure—The weight of the air surrounding the earth in layers of varying characteristics. Standard atmospheric pressure is expressed as 29.92 inches of mercury or 1613.2 millibars.

avionics—A catch-all phrase for communication, navigation and related instrumentation in an aircraft. A contraction of *aviation electronics.*

balanced field length—The distance within which a jet aircraft can accelerate to V_1 and then either stop or accelerate to a safe climb speed, V_2, and clear a 35-foot obstacle on one engine.

base leg—A part of the airport traffic pattern. A flight path at right angles to the runway, following the downwind leg and followed by the final approach.

C.—Centigrade or Celsius.

calibrated airspeed—Indicated airspeed corrected for instrument and installation errors.

CAS—Calibrated airspeed.

carburetor heat—A heating unit located near the carburetor throat and controlled by a plunger in the cockpit. It is used to melt carburetor ice.

carburetor ice—Ice forming in the carburetor throat due to excessive moisture in the air.

CDI—Course-deviation indicator.

com—Communication (radios).

compass, gyro—A compass system driven by a gyroscope and not reacting to the magnetic field of the earth.

compass, magnetic—A compass which, during straight and level flight, automatically aligns itself with the magnetic north.

compass rose—The compass card showing the 360 degrees.

constant-speed propeller—A controllable-pitch propeller which maintains a constant rpm by automatically changing the blade angle in relation to engine output.

course—The direction of flight of an aircraft across the ground.

course deviation indicator—The needle, bar, or other indicator in an OBI which displays the position of an aircraft relative to a radial or bearing from or to a VOR.

CRS—Course.

dead reckoning—A method of navigation by which the course and time an aircraft between two given points is estimated by taking course, speed, and wind components calculated with a wind triangle into consideration. The phrase *dead* is a bastardization of the term *deduced* reckoning.

ded reckoning—Dead reckoning.

density altitude—Pressure altitude corrected for prevailing temperature conditions.

detonation—The burning of fuel-air mixture by explosion rather than steady burning. It results in rapidly rising cylinder-head temperatures and can result in severe engine damage.

dew point—The temperature to which air must cool in order for condensation to take place without change in pressure or vapor content.

DG—Directional gyro (compass).

DH—Decision height.

directional gyro—A gyroscopic flight instrument which, when set to conform with the magnetic compass, will continue to indicate the aircraft heading for some time, regardless of turns or pitch changes. It tends to develop heading errors and must be adjusted intermittantly.

distance measuring equipment—A combination of airborne and ground equipment which gives a reading of the distance from the airplane to a ground station by measuring the time lapse of a signal generated in the airplane and reflected by the ground station.

DME—Distance measuring equipment.

downwind—Away from the direction from which the wind is blowing.

downwind leg—The flight path parallel to the runway in the direction opposite to landing. It is part of the standard airport traffic pattern.

drag—The force created by friction of the air on objects in motion. It must be overcome by thrust in order to achieve flight parallel to the relative wind. There are two types of drag, *induced* drag and *parasite* drag. Induced drag is created through the process of creating lift. Parasite drag is all drag from surfaces which do not contribute to lift. It increases with an increase in airspeed.

EGT—Exhaust-gas temperature (gauge).

elevator—The primary control surface attached to the horizontal stabilizer, which can be moved up or down to control the pitch of the aircraft. It is, in fact a speed control, not an altitude control.

encoding altimeter—An instrument which senses the current altimeter reading and gives this information to the transponder which, in turn, automatically transmits it to ATC.

engine analyzer—An exhaust-gas-temperature gauge with probes and readouts for each cylinder of the engine.

E6b—A circular sliderule-type computer used to compute a variety of flight-related mathematic problems.

exhaust-gas temperature—The temperature of the gases escaping through the exhaust manifold. The temperature of these gases varies with the fuel-air mixture being used and it can be measured and displayed in the cockpit through the use of an exhaust-gas-temperature gauge or an engine analyzer.

F —Fahrenheit.

FAA—Federal Aviation Administration.

fanjet—An aircraft driven by turbine engine combining the effects of direct jet thrust with the thrust produced by an engine-driven shrouded fan.

final approach—The final portion of an airport traffic pattern during which the aircraft is aligned with the runway centerline.

fixed-pitch propeller—A propeller, the blade angle of which cannot be changed or adjusted.

FL—Flight level; FL180 stands for 18,000 feet.

flaps—Auxiliary control surfaces, usually located at the trailing edges of the inner wing panels, between the fuselage and the ailerons. Flaps can be extended and/or turned down to increase the wing camber and/or surface, creating additional lift and drag.

flare—A smooth leveling of the aircraft during which the nose is raised at the end of the glide and just prior to touchdown.

fpm—Feet per minute, in terms of rate of climb or descent.

ft—Feet.

GCA—Ground-controlled approach. A non-precision approach during which a ground-based controller uses radar to provide guidance for the pilot.

generator—A device, identical in construction to an electric motor which, when driven by the engine, generates electrical current and continuously recharges the battery.

gph—Gallons per hour.

ground control—An ATC service at controlled airports, responsible for the safe and efficient movement of aircraft and airport vehicles on the ground.

ground effect—A certain amount of additional lift which takes effect when the aircraft is close to the ground. It is the result of air being compressed between the wings and the ground, and low-wing aircraft are more susceptible to the effect than are high-wing aircraft.

ground speed—The speed with which an aircraft moves across the ground.

GS—Ground speed.

HDG—Heading.

heading—The direction in which the aircraft flies through the air, not with reference to the ground. In other words, the direction in which the nose of the aircraft is pointing.

horizontal stabilizer—The fixed horizontal section of the empennage to which the elevator is attached.

IAS—Indicated airspeed.

IFR—Instrument flight rules.

ILS—Instrument landing system.

IM—Inner marker.

inches of mercury—A unit of measurement of atmospheric pressure, indicating the height in inches to which a column of mercury will rise in a glass tube in response to the weight of the atmosphere exerting pressure on a bowl of mercury at the base of the tube.

indicated airspeed—The airspeed which is shown by the airspeed indicator. It is nearly always less than true airspeed, but usually not much different from calibrated airspeed.

induced drag—See *drag*.

in hg—Inches of mercury.

intersection takeoff—A takeoff using less than the full length of the runway.

kHz—KiloHertz or kilocycles.

knots—Nautical miles per hour.

kts—Knots.

laminar flow—Air flowing smoothly over and adhering to the surface of an airfoil.

lift—The generally upward force created by the difference of pressure between the upper and lower surfaces of an airfoil in motion. In level flight it is balanced by the force of gravity.

LOC—Localizer.

magnetic course—The course of an aircraft referenced to the magnetic north.

magnetic heading—The heading of an aircraft referenced to the magnetic north.

magnetic north—The region, some distance from the geographic north pole, where the earth's magnetic lines concentrate.

magneto—A self-contained generator which supplies electrical current to the spark plugs in the ignition system.

manifold—An arrangement of tubing with one orifice on one end and several on the other.

manifold pressure—The pressure of the fuel-air mixture in the intake manifold.

MAP—Missed-approach point.

MDA—Minimum descent altitude.

MEA—Minimum en-route altitude.

MHz—MegaHertz or megacycles.

mixture—The mixture of fuel and air necessary for combustion in reciprocating engines.

mls—Height in feet above mean sea level.

MM—Middle marker.

MOA—Military operations area.

MOCA—Minimum obstruction-clearance altitude.

monocoque—Type of all-metal aircraft construction in which the fuselage shell carries most of the structural loads.

MP—Manifold pressure.

mph—Statute miles per hour.

MRA—Minimum reception altitude.

mushing—Flying along in a nose-high attitude at low speed.

NASA—National Aviation and Space Administration.

nav—Navigation (radio).

NDB—Non-directional beacon.

needle and ball—An instrument which shows the degree of bank of the aircraft and displays whether the aircraft is in a skid or a slip. An older version of the more modern turn-and-bank indicator.

nm—Nautical mile(s).

normally aspirated—A reciprocating engine which is not turbocharged or mechanically supercharged.

OAT—Outside-air temperature (gauge).

OBI—Omni-bearing indicator.

OBS—Omni-bearing selector.

OM—Outer marker.

PAR—Precision approach radar (approach). A ground controlled approach in which the controller uses radar which displays the position of the aircraft both horizontally and vertically.

parasite drag—See *drag*.

pattern—Airport landing pattern; the downwind and base leg and the final approach.

peak—Expression used to denote the highest possible exhaust-gas temperature.

pitch—The attitude of the aircraft with reference to a horizontal axis at right angles to the fuselage. In other words, nose-down or nose-up.

pilotage—Navigation by reference to visible landmarks. Used usually in conjunction with Sectional charts on which all meaningful landmarks are shown.

pitot—static system—A device which compares pitot pressure with static or atmospheric pressure and presents the result in the cockpit by means of the airspeed indicator, the altimeter, and the vertical-speed indicator.

pitot tube—A protrusion with a small orifice, exposed to the airstream and designed to measure the pressure with which an aircraft meets the air.

pph—Pounds per hour.

precession—The tendency of a directional gyro to gradually become unreliable due to friction.

preignition—The burning of the fuel-air mixture in the combustion chamber before the spark plugs have had an opportunity to fire. It usually follows excessive overheating of the engine, and can result in serious engine damage.

propeller—A device consisting of two or more airfoil-shaped blades which is designed to convert the turning force of the engine into thrust.

propjet—A jet aircraft in which the turbine engine drives a propeller.

psi—Pressure in terms of pounds per square inch.

relative wind—The movement of air relative to the movement of an airfoil. It is parallel to and in the opposite direction of the flight path of an airplane.

RNAV—Area navigation.

rmp—Revolutions per minute.

rudder—The primary control surface attached to the vertical stabilizer, movement of which causes the tail of the aircraft to swing either left or right. It controls yaw.

runup—A pre-takeoff check of the performance of the engine and, in aircraft equipped with constant-speed props, the operation of the propeller.

RVR—Runway visual range; a measurement of airport visibility expressed in feet.

SAR—Surveillance radar approach. See *GCA*.

scan—The consecutive order in which the pilot checks the instrument readings in the cockpit.

sectional chart—An aeronautical chart of a section of the U.S. at a scale of 1:500,000 or approximately seven nm per inch.

service ceiling—The highest altitude at which a given aircraft can continue to climb at 100 fpm.

skid—Lateral movement of an airplane toward the outside of a turn, caused by incorrect use of the rudder.

slip—The tendency of an aircraft to lose altitude by slipping toward the center of a turn as a result of incorrect use of the rudder.

sm—Statute mile(s).

spin—A maneuver in which the airplane, after stalling, descends nearly vertically, nose-low, with the tail revolving around the vertical axis of the descent.

spoilers—Flat vertical surfaces which can be raised out of the upper surface of the wing at the pilot's discretion. They spoil the airflow over the wing and cause it to lose lift and therefore drop. Spoilers may be constructed to be deployed simultaneously in both wings or individually in one wing or the other. In the first instance they act as speed brakes. In the latter they take the place of ailerons.

stall—The inability of an airplane to continue flight due to an excessive angle of attack. It will either drop its nose and thus reduce the angle of attack and regain flying speed or, if forced to retain the excessive angle of attack, it may fall into a spin.

stall speed—The speed at a given angle of attack at which airflow separation begins and the stall occurs. Aircraft can stall at virtually any speed if an acceptable angle of attack is exceeded.

stall-spin—The combination of a stall followed by a spin, a major cause of fatal accidents.

stall warning—A device, usually involving a buzzer, a light or both, which indicates to the pilot that the aircraft is about to stall.

static vent—A hole, usually located in the side of the fuselage, which provides air at atmospheric pressure to operate the pitot-static system.

stick—Control wheel or yoke.

super-critical wing—A term used for an airfoil design created by NASA with the help of computers. It has low drag and high lift characteristics.

S/VFR—Special VFR.

TAS—True airspeed.

taxi—Moving an aircraft on the ground under its own power.

TBO—Time between major overhauls, expressed in hours.

TCA—Terminal control area.

thrust—The forward force, pushing or pulling, exerted by the engine and, in turn, by the propeller or, in the case of gliders, by gravity. It opposes and must overcome drag.

torque—The normal tendency of an aircraft to rotate to the left in reaction to the right-hand rotation of the propeller and the fact that the action of the propeller forces air against the left side of the vertical stabilizer with greater force than against the right side. It alters with changes in power.

tower—Control tower at a controlled airport.

transponder—An airborne radar-beacon transceiver that automatically transmits responses to interrogation by ground-based transmitters.

trim tab—A small airfoil attached to the control surface, primarily the elevator and occasionally the rudder, which can be adjusted to cause changes in the position of the control surface under varying flight conditions.

true airspeed—The actual speed at which an aircraft is moving through the air. It is calibrated airspeed adjusted for actual air density and altitude.

true course—Course referenced on true north.

true heading—Heading referenced to true north.

true north—The geographic (not magnetic) north. The direction of the north pole from any point on the globe.

T-tail—The tail section of an airplane where the horizontal stabilizer or stabilator is attached to the top of the vertical stabilizer.

turbocharging—A process in which a turbine, driven by the exhaust gases, compresses the air and thus increases the amount of fuel-air mixture available to the engine.

turbojet—A jet aircraft propelled by jet engines without the addition of fans or propellers.

turboprop—See *propjet*.

turn-and-bank indicator—See *needle and ball*.

unicom—Aeronautical advisory station for communication with aircraft (and the associated frequencies). Unicoms are usually manned by FBOs or airport personnel and provide pilots with such information as the active runway, wind direction and velocity, and other conditions of importance to the pilot. Unicoms are not authorized to give takeoff or landing clearances, or in any way control traffic (except when relaying word from ATC, in which case any such transmission must be preceded by *ATC clears . . .*).

unusual attitude—Any attitude of an aircraft in terms of pitch or roll or both which is beyond the normal operating attitude. Recovery from unusual attitudes by reference to instruments is an important part of instrument training.

venturi—A tube which is narrower in the middle than at either end. When air is forced through a venturi it results in suction which can be used to drive gyro instruments on aircraft without vacuum pumps.

vertical-speed indicator—An instrument, part of the pitot-static system, which indicates the rate of climb or descent in terms of fpm. It is usually calibrated in units of either 100 or 1,000 fpm.

vertical stabilizer—A fixed vertical airfoil on the empennage to which the rudder is attached.

VFR—Visual flight rules.

VHF—Very high frequency; electromagnetic frequencies between 30 and 300 MHz.

VOR—Very high frequency omni-directional radio range; A ground-based VHF navigation aid.

VOR/DME—A VOR with DME capability.

VORTAC—Another type of VOR with DME capability.

VSI—Vertical-speed indicator.

WACs—World aeronautical charts. Scale: 1:1,000,000.

wake turbulence—Turbulence created by the movement of an aircraft through the air. Primarily the trailing wingtip vortices which develop in the wake of heavy aircraft. Resembling a pair of counter-rotating horizontal tornadoes, they are at their worst behind slow-moving heavy jet aircraft just after takeoff and immediately prior to landing. (All aircraft develop wake turbulence to some degree, but with lighter aircraft it is of no great consequence.) To avoid getting caught in wake turbulence, light aircraft should stay above the flight path of the preceding heavy aircraft and land *beyond* the point at which that aircraft's wheels touched down. On takeoff a light aircraft should lift off before the point at which the preceding aircraft lifted off, and then stay above and, if possible, upwind of that aircraft's flight path.

waypoint—A navigational fix used in area navigation, created by electronically relocating a VOR/DME or VORTAC from its actual position to a position desired by the pilot.

wind shear—An abrupt change in wind direction or velocity.

wing leveler—A simple autopilot without directional control capability.

yaw—The movement of an aircraft to either side, turning around its vertical axis, without banking.

yoke—Control wheel; stick.

zulu time—Greenwich mean time.

Index